Top 20 Essential Skills for ArcGIS® Utility Network

TOP 20
ESSENTIAL SKILLS FOR ARCGIS® UTILITY NETWORK

Melissa L. Mayo, GISP
Christopher Cremons
Remi Myers

Esri Press
Redlands, California

Esri Press, 380 New York Street, Redlands, California 92373-8100
Copyright © 2025 Esri
All rights reserved.
Printed in the United States of America.

ISBN: 9781589488274
Library of Congress Control Number: 2025931580

The information contained in this document is the exclusive property of Esri or its licensors. This work is protected under United States copyright law and other international copyright treaties and conventions. No part of this work may be reproduced or transmitted in any form or by any means, electronic or mechanical, including photocopying and recording, or by any information storage or retrieval system, except as expressly permitted in writing by Esri. All requests should be sent to Attention: Director, Contracts and Legal Department, Esri, 380 New York Street, Redlands, California 92373-8100, USA.

The information contained in this document is subject to change without notice.

US Government Restricted/Limited Rights: Any software, documentation, and/or data delivered hereunder is subject to the terms of the License Agreement. The commercial license rights in the License Agreement strictly govern Licensee's use, reproduction, or disclosure of the software, data, and documentation. In no event shall the US Government acquire greater than RESTRICTED/LIMITED RIGHTS. At a minimum, use, duplication, or disclosure by the US Government is subject to restrictions as set forth in FAR §52.227-14 Alternates I, II, and III (DEC 2007); FAR §52.227-19(b) (DEC 2007) and/or FAR §12.211/12.212 (Commercial Technical Data/Computer Software); and DFARS §252.227-7015 (DEC 2011) (Technical Data–Commercial Items) and/or DFARS §227.7202 (Commercial Computer Software and Commercial Computer Software Documentation), as applicable. Contractor/Manufacturer is Esri, 380 New York Street, Redlands, California 92373-8100, USA.

Esri products or services referenced in this publication are trademarks, service marks, or registered marks of Esri in the United States, the European Community, or certain other jurisdictions. To learn more about Esri marks, go to: links.esri.com/EsriProductNamingGuide. Other companies and products or services mentioned herein may be trademarks, service marks, or registered marks of their respective mark owners.

For purchasing and distribution options (both domestic and international), please visit esripress.esri.com.

CONTENTS

Introduction	ix
Acknowledgments	xi
Getting started	xiii
How to use this book	xix

1 Exploring ArcGIS® Pro for ArcGIS Utility Network — 1

Introduction	1
Tutorial 1-1: Walking through ArcGIS Pro	2
Tutorial 1-2: Exploring ArcGIS Utility Network geoprocessing tools	5
Take the next step	6
Summary	6

2 Basic editing in ArcGIS Utility Network — 7

Introduction	7
Tutorial 2-1: Walking through the basics of editing	8
Take the next step	12
Summary	13

3 Creating feature templates — 14

Introduction	14
Tutorial 3-1: Building feature templates using group templates	15
Tutorial 3-2: Building feature templates using preset templates from selected features	20
Take the next step	24
Summary	24

4 Overview of structure and domain networks with tiers — 25

Introduction	25
Tutorial 4-1: Exploring the structure network	26
Tutorial 4-2: Working with domain networks	31
Tutorial 4-3: Examining tier definitions	36
Take the next step	37
Summary	37

5 Applying feature subtypes using asset groups and asset types — 38

Introduction — 38
Tutorial 5-1: Introduction to subtypes — 39
Tutorial 5-2: Configuring default asset types — 42
Take the next step — 44
Summary — 44

6 Terminals and controllers — 45

Introduction — 45
Tutorial 6-1: Configuring and setting terminals on a device — 46
Tutorial 6-2: Set a subnetwork controller — 49
Take the next step — 52
Summary — 52

7 Using network attributes and categories — 53

Introduction — 53
Tutorial 7-1: Exploring properties and management of network categories — 54
Tutorial 7-2: Exploring the composition and impacts of network attributes — 55
Take the next step — 59
Summary — 59

8 Applying network rules and validating data — 60

Introduction — 60
Tutorial 8-1: Examining different types of network rules — 61
Tutorial 8-2: Working with network rules in editing — 65
Take the next step — 67
Summary — 67

9 Using the error inspector to resolve error features — 68

Introduction — 68
Tutorial 9-1: Exploring the error inspector — 69
Take the next step — 73
Summary — 73

10 Creating trace locations — 74

Introduction — 74
Tutorial 10-1: Creating starting points — 75
Tutorial 10-2: Creating barriers — 78
Take the next step — 79
Summary — 80

Contents vii

11 Using a basic connected trace — 81
Introduction — 81
Tutorial 11-1: Configuring and running a connected trace — 82
Take the next step — 85
Summary — 85

12 Using directional traces — 86
Introduction — 86
Tutorial 12-1: Working with a downstream trace — 87
Tutorial 12-2: Working with an upstream trace — 88
Take the next step — 90
Summary — 90

13 Applying function barriers — 91
Introduction — 91
Tutorial 13-1: Understanding and creating function barriers — 92
Tutorial 13-2: Understanding and creating filter function barriers — 94
Take the next step — 97
Summary — 98

14 Using condition and filter barriers — 99
Introduction — 99
Tutorial 14-1: Creating and accessing a condition barrier — 100
Tutorial 14-2: Creating a filter barrier — 105
Take the next step — 106
Summary — 107

15 Working with functions — 108
Introduction — 108
Tutorial 15-1: Creating a function — 109
Take the next step — 113
Summary — 113

16 Working with trace output configurations — 114
Introduction — 114
Tutorial 16-1: Using output asset types — 115
Tutorial 16-2: Using output conditions — 117
Tutorial 16-3: Using result types — 119
Take the next step — 124
Summary — 124

17 Working with network diagrams — 125

Introduction — 125
Tutorial 17-1: Creating a network diagram from a traced subnetwork — 126
Take the next step — 130
Summary — 130

18 Modifying diagrams with layout configurations — 131

Introduction — 131
Tutorial 18-1: Applying a layout to a network diagram — 132
Take the next step — 138
Summary — 138

19 Configuring network controllers — 139

Introduction — 139
Tutorial 19-1: Evaluating the subnetwork controller definition and creating a controller — 140
Take the next step — 143
Summary — 144

20 Managing subnetworks — 145

Introduction — 145
Tutorial 20-1: Updating and managing a subnetwork — 146
Take the next step — 153
Summary — 153

Conclusion — 154
Glossary — 155

INTRODUCTION

ArcGIS® Utility Network is flexible and easily adaptable to wires, pipes, fiber, and other infrastructure while augmenting additional decision support systems commonly used by utilities. This flexibility enables users to fully model and analyze utility systems while incorporating data from external sources. Although more and more resources have become available to support and train users on this product, the time has come for a publication highlighting the most important skills necessary to successfully use Utility Network as a GIS technician or analyst.

Top 20 Essential Skills for ArcGIS Utility Network has been broken down into four subject areas: overview, data structure, trace framework, and network management. We explore the components and elements of the utility network as a user for our community, but our primary goal was to provide a desktop companion that would ease users' transition and allow them to take full advantage of its capabilities.

The tutorials in this book use ArcGIS Solutions foundation packages, which provide a standard set of data for electric, water, and gas industries. It's our intent that by using these datasets, readers will be able to walk through the utility network in an easier, stress-free fashion. A variety of industry types are referenced throughout the book, showing use cases for the product across industries. We recommend that readers review the "Getting Started" section of this book before starting any of the chapters. This step is essential for users to best experience the technology and methodology covered in this book. We selected these 20 skills based on user input, suggestions, and frustrations over the last few years in an effort to address them. Moving forward, we hope to continue collecting feedback from the user community, further improving our communication regarding the use of this emerging technology to best support our communities.

ACKNOWLEDGMENTS

We would be remiss if we failed to recognize all the developers and engineers who have labored over ArcGIS Utility Network with the hope of giving our customers the best experience possible. This content relies on the feedback from trainers and implementers who help our users adopt Utility Network in their organizations.

Our earnest appreciation goes out to our customers, partners, and distributors across the utility, infrastructure, and GIS communities for pushing us to release this material. It's always a pleasure to engage with you, and your input is critical to moving geospatial technology forward. We love seeing you at conferences, on webinars, and in our day-to-day engagements; your efforts, products, and stories are an inspiration, and we all truly appreciate everything you do to make our world a better place.

Special thanks go out to our reviewers, who encouraged and supported us as we crossed the publication finish line. They helped us make certain that we provided the best content possible in a format that would resonate with our readers. Special thanks and appreciation go to Jon DeRose, Mohana Krishna Punnam, and Heath Birchfield, who were especially helpful during the editing process.

Finally, we can't forget to mention Legend, the floofy map dog, without whose consistent support and encouragement we could not have completed this effort. Woof!

GETTING STARTED

In this preliminary chapter, you'll explore the three Foundation datasets in ArcGIS Solutions that are used with the tutorials in this book. All the compressed folders have a few elements in common: a utility network, a data dictionary, geoprocessing tools, and an ArcGIS Pro project. We'll explore the various packages.

- ElectricUtilityNetworkFoundationV2_2
- GasPipelineReferencingUtilityNetworkFoundationV2_1
- Water_Distribution_Utility_Network_Foundation_v1_2

Tutorial GS-1: Exploring the Electric Utility Network Foundation

We'll start with the Electric Utility Network Foundation.

1. Open the **ElectricUtilityNetworkFoundationV2_2** folder.

   Inside the uncompressed **ElectricUtilityNetworkFoundationV2_2** folder, the **ElectricUtilityNetworkFoundation** ArcGIS Pro project file (.aprx) is your primary workspace for the tutorials.

   - Data Dictionary
   - Data Loading Tools
   - Database
   - ElectricUtilityNetworkFoundation.gdb
   - Styles
   - Toolboxes
   - ElectricUtilityNetworkFoundation.aprx
   - ElectricUtilityNetworkFoundation.tbx

Set up the project space

2. Double-click **ElectricUtilityNetworkFoundation.aprx** to open the project in ArcGIS Pro.

   Depending on the version of ArcGIS Pro that you have, there may be a bit of upgrade time when starting the project for the first time. When you open the project file, the **Electric Network Editor** map appears.

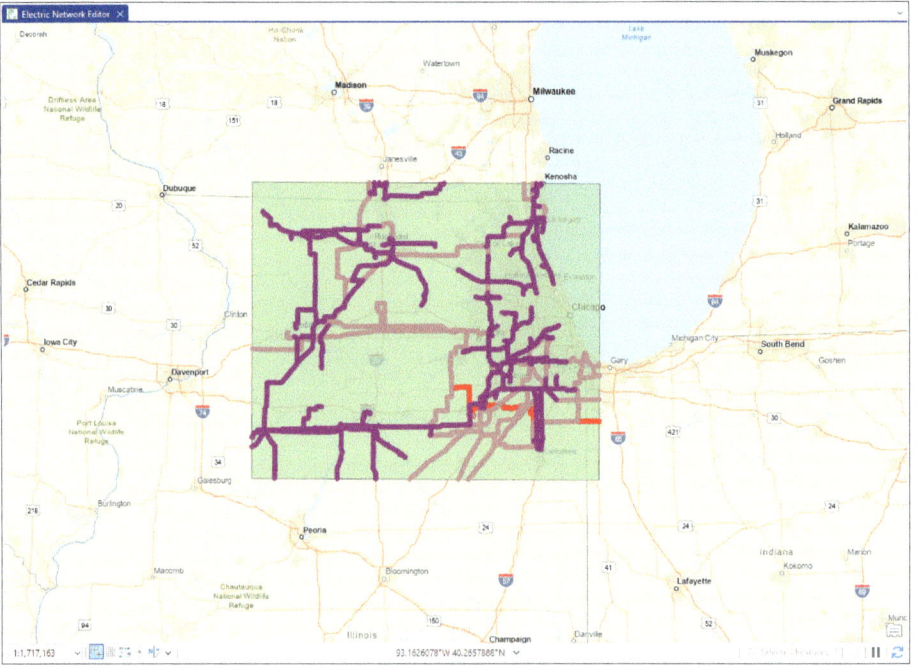

   You'll activate the **Contents** and **Catalog** panes so that they're visible in your ArcGIS Pro project.

3. On the ribbon, click the **View** tab. In the **Windows** group, click **Catalog Pane** and **Contents**.

   This map is your starting point for the tutorials using the **Electric Utility Network Foundation** dataset.

Tutorial GS-2: Exploring the Gas Pipeline Utility Network Foundation

Let's move on to the **Gas Pipeline Utility Network Foundation** dataset.

1. Open the **GasPipelineReferencingUtilityNetworkFoundationV2_1** folder.

   The uncompressed **GasPipelineReferencingUtilityNetworkFoundationV2_1** folder contains the **Gas and Pipeline Utility Network Foundation** ArcGIS Pro project file, which is your primary workspace for the tutorials.

   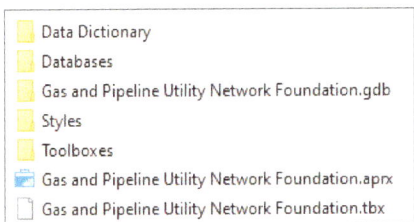

Set up the project space

2. Double-click **ElectricUtilityNetworkFoundation.aprx** to open the project in ArcGIS Pro.

   The **Gas and Pipeline Network Editor** map opens in the map layout.

   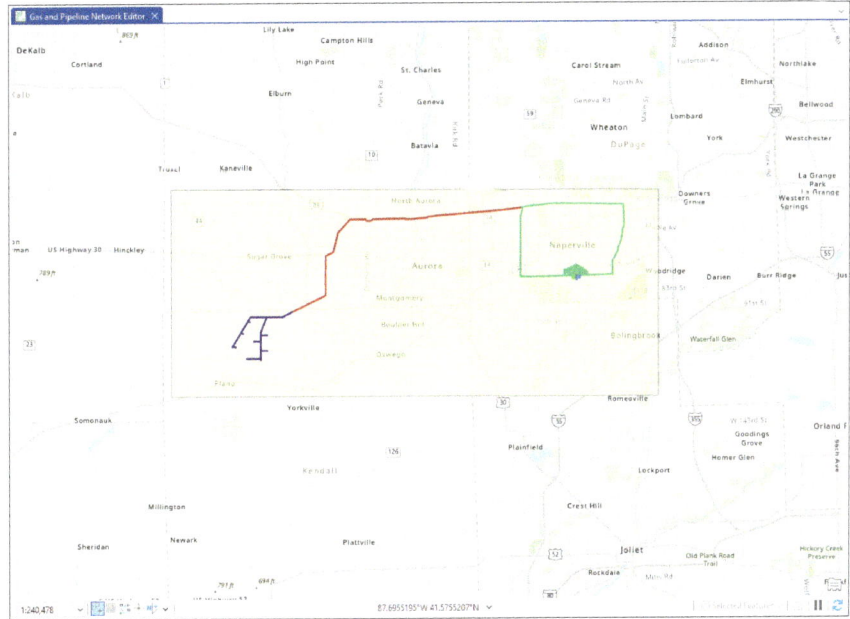

3. Repeat the steps from tutorial GS-1 to open the **Catalog** pane and the **Contents** pane.

This is your starting point for the tutorials using the **Gas and Pipeline Utility Network Foundation** dataset.

Tutorial GS-3: Exploring the Water Distribution Foundation

Let's move on to the **Water Distribution Utility Network Foundation** dataset.

Inside the uncompressed **Water_Distribution_Utility_Network_Foundation_v1_2** folder, the **Water Distribution Utility Network Foundation** ArcGIS Pro project file is your primary workspace for the tutorials.

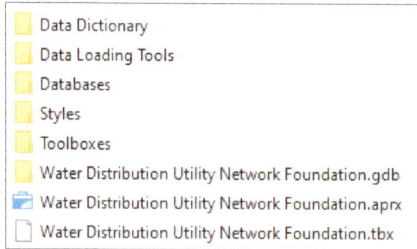

Set up the project space

1. Double-click **Water Distribution Utility Network Foundation.aprx** to open the project in ArcGIS Pro.

   The **Gas and Pipeline Network Editor** map opens in the map layout.

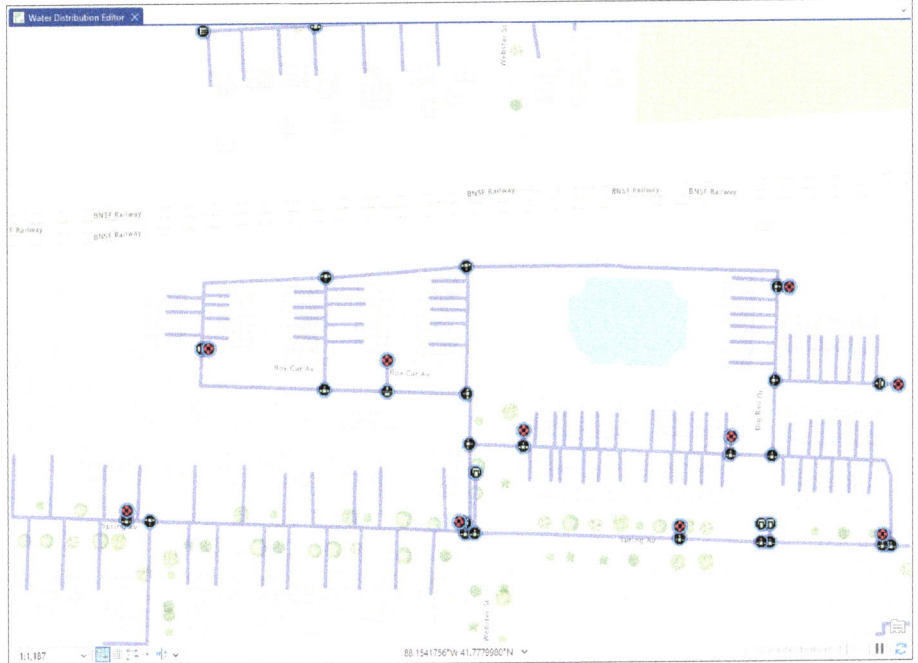

2. Repeat the steps from tutorial GS-1 and tutorial GS-2 to open the **Catalog** pane and the **Contents** pane.

This is your starting point for the tutorials using the **Water Distribution Utility Network Foundation** dataset.

On your own

Explore the layers in the ArcGIS Solutions Foundation datasets.

Take the next step

The ArcGIS Foundation datasets are interchangeable, and all the tutorials can be used across any of the ArcGIS Utility Network datasets provided by the ArcGIS Solutions team.

Summary

The ArcGIS Foundation datasets were designed to enable customers and partners to learn more about ArcGIS Utility Network capabilities. These packages contain all the basic components to explore the utility network and extend network capabilities through prototypes and configurable datasets.

Workflow

1. Configure the Electric Utility Network Foundation dataset for tutorials.
2. Configure the Gas and Pipeline Utility Network Foundation dataset for tutorials.
3. Configure the Water Distribution Utility Network Foundation dataset for tutorials.

HOW TO USE THIS BOOK

About this book

Top 20 Essential Skills for ArcGIS Utility Network has been tested for compatibility with ArcGIS Pro 3.3.

This book is designed for users who want to learn ArcGIS Utility Network on their own. Readers should be familiar with ArcGIS Pro and using file geodatabases. Readers need no prior ArcGIS Utility Network knowledge to complete this book. Each chapter uses one or more tutorials to demonstrate the related skill in a hands-on environment and should take about 45 minutes to complete.

Read all the text and take your time. Avoid lightly scanning the instructions or clicking without knowing why. Read the explanations.

The "Take the Next Step" section at the conclusion of each tutorial is optional but recommended. These sections provide additional suggestions, add functionality, and further refine the project you've completed. While the workflow is fresh in your mind, this section is your chance to solidify what you've learned and further develop your skills.

The "Workflow" section at the end of a chapter provides a simplified version of what you learned. If you've completed the chapter, the workflow should help you repeat the workflow with other data, whenever you need.

Hardware and software requirements

To perform the tutorials in this book, it's recommended that you use ArcGIS Pro 3.3 on a computer that's running the Windows operating system. But earlier software versions may be used if needed. Hardware requirements for ArcGIS Pro are available at links.esri.com/SysReqs.

Licensing the software

If you have existing ArcGIS credentials or can obtain credentials from your educational institution or organization to access the required software listed in this section, you may use those credentials and proceed. Regarding Utility Network, these tutorials are built on a file geodatabase and no additional extensions are needed.

Information about software trial options as well as Personal Use and Student Use licensing can be found on esri.com.

Downloading the tutorial data

The tutorial data for this book is available on ArcGIS Online at links.esri.com/UN20-Data.

On your computer, create a folder in your C drive called Top20ArcGISUtilityNetwork. Download the tutorial data and store it in this folder: C:\Top20ArcGISUtilityNetwork.

Tutorial data that accompanies this book is covered by a license agreement that stipulates the terms of use. You can read the license agreement at links.esri.com/LicenseAgreement.

Resources and learning

ArcGIS Utility Network resources

Go to links.esri.com/ArcGISUNResources for a variety of resources:

- Read blogs about new capabilities and features written by the ArcGIS Utility Network team.
- Watch video tutorials on the implementation and use of Utility Network for all industries.
- Register for instructor-led and self-paced training on using Utility Network in your organization.

ArcGIS Utility Network documentation

Visit links.esri.com/ArcGISUNDocumentation for Esri documentation and Help topics that can answer your questions and provide in-depth details and troubleshooting tips.

Esri Community

Visit links.esri.com/ArcGISUNCommunity to post questions, share ideas, and engage with other ArcGIS Utility Network users.

Feedback and updates

For feedback, updates, or collaboration, visit the Esri Press page on Esri Community, where users can ask questions and share experiences. Access it at links.esri.com/EsriPressCommunity.

Additional information is available on the book's web page at links.esri.com/UN20.

CHAPTER 1
Exploring ArcGIS® Pro for ArcGIS Utility Network

Objectives

- Describe how to configure ArcGIS Pro to best use ArcGIS Utility Network.
- Locate key functions for ArcGIS Utility Network.
- Locate key geoprocessing tools for ArcGIS Utility Network.

Introduction

Although ArcGIS Utility Network was intended to be used across desktop, web, and mobile solutions, the tutorials covered in this book will use ArcGIS Pro. The purpose of this chapter is to review key ArcGIS Pro functionality as it pertains to Utility Network. This chapter features the Water Distribution Utility Network Foundation for the tutorials with the Water Distribution Editor map. The following image shows the pipes, devices, assemblies, and other features used to represent a water network.

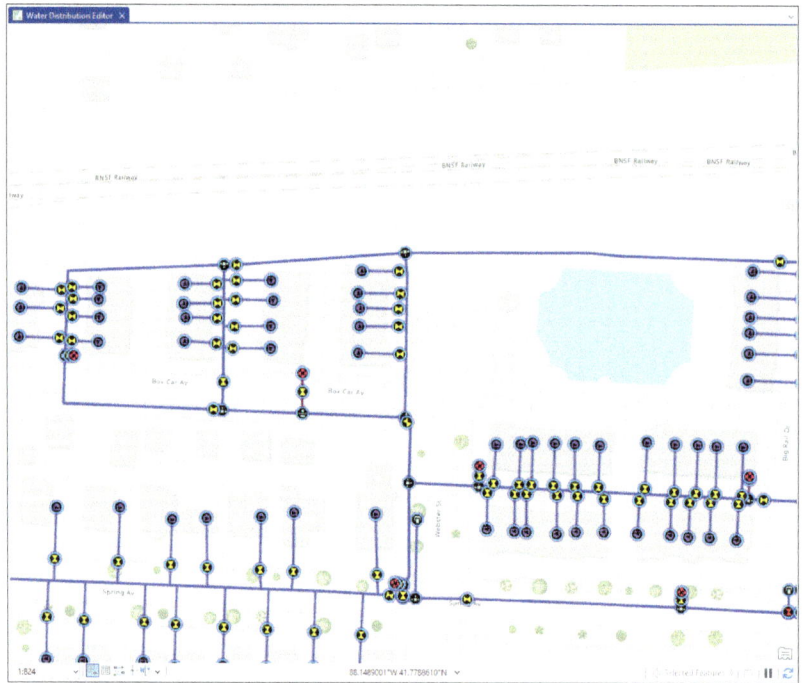

Tip: For information about how to set up your utility network project, see the "Getting Started" section in the front of the book. The tutorials in this chapter can be used with any Network Foundation dataset.

Tutorial 1-1: Walking through ArcGIS Pro

1. From the **Datasets_For_UN_Skills_Book** folder, open **Water Distribution Utility Network Foundation.aprx**.

Explore the tabs and ribbons

In ArcGIS Pro, tabs have a range of map tools and functions that are clustered by functionality. In this tutorial, you'll start with the **Windows** group of the **View** tab to open the **Catalog** pane.

2. At the top of the screen, on the ribbon, click the **View** tab.

3. In the **Windows** group, click the **Catalog Pane** button.

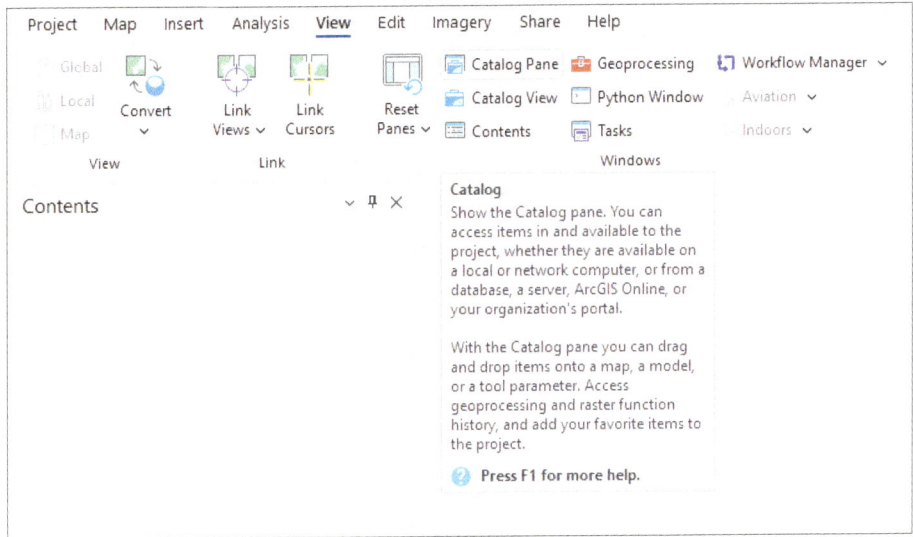

Examine the Catalog pane

4. With the **Catalog pane** open, expand the **Maps** folder to see the preconfigured map in the **Foundation Package**.

In addition to **Maps**, there are also tabs for local folders specific to the project, such as **Toolboxes** with geoprocessing tools, **Databases**, **Styles**, and other elements.

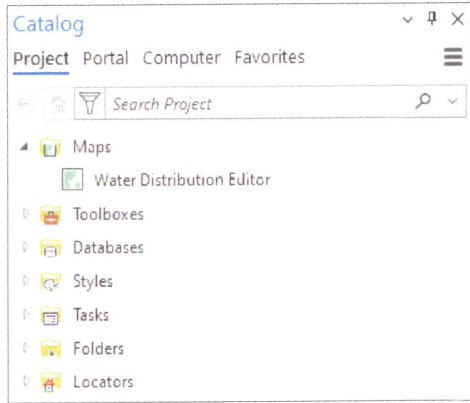

Open the Water Distribution Editor Map

5. In the **Maps** folder, double-click the **Water Distribution Editor** map.

   The map opens with an active instance of ArcGIS Utility Network.

6. Return to the **View** tab. In the **Windows** group, click the **Contents** button to enable the **Contents** pane for visualizing layers within the map.

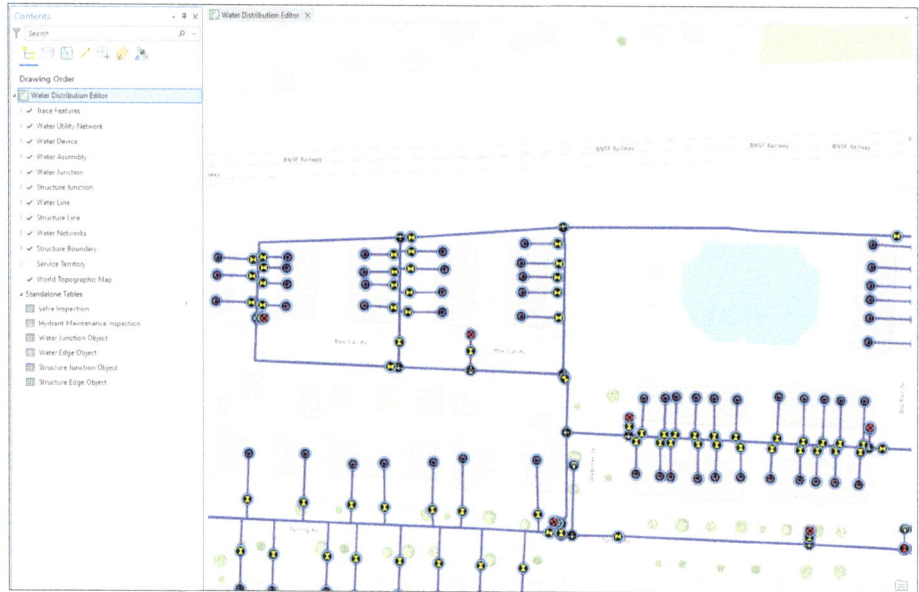

Looking at the ribbon across the top of the map project, notice that the **Utility Network** tab is now showing and available to use.

Explore the Utility Network tab

Now that an active utility network is opened in the map view, you'll look closely at the **Utility Network** tab.

7. Click the **Utility Network** tab.

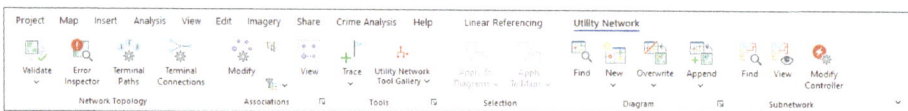

Starting from the left side of the tab, the **Network Topology** group is where tools are found to test features for correctness, fix features with errors, and configure devices and lines that connect through terminals. The next group, **Associations**, provides tools to help users manage feature associations. The **Tools** group has a range of tools that establish trace properties and functions. The **Selection** group components enable selected features to be propagated to other map or diagram elements. The **Diagram** group enables users to explore, build, and modify schematics generated from the network. Finally, the **Subnetwork** group is provided to help users configure their ability to manage circuits or feeders.

Tutorial 1-2: Exploring ArcGIS Utility Network geoprocessing tools

Open the Geoprocessing pane

1. The **Geoprocessing** toolbox is used to store and make scripts and tools available through an accessible user interface (UI). These scripts and tools are organized by core functions and exposed through licensing capabilities. For Utility Network, a range of administrative and user experience tools affect the network index, trace, and even network diagrams.

2. On the ribbon, click the **Analysis** tab.

3. In the **Geoprocessing** group, click **Tools** to access the **Geoprocessing** pane.

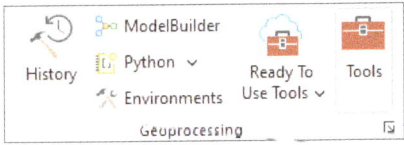

Search for Utility Network tools

In the **Geoprocessing** pane, you can find **Utility Network** tools by using the search dialog box or scrolling down to the **Utility Network Tools** toolbox.

4. Locate the **Utility Network Tools** toolbox.

    These tools, such as **Validate Network Topology** and **Trace**, are duplicates of tools displayed on the **Utility Network** tab.

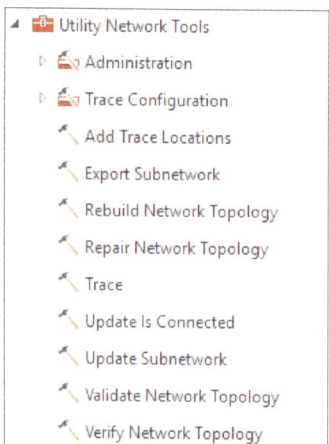

On your own

As you work through the chapters in the book, explore how tools can be located in the Geoprocessing toolkit.

Take the next step

This chapter provided a brief overview of ArcGIS Pro using an ArcGIS Utility Network dataset. Many of the tools provide help guides (indicated by a blue question mark) to show how features and functions can be used.

Summary

ArcGIS Pro is a powerful application for demonstrating the full capabilities of ArcGIS Utility Network. It provides numerous techniques and approaches for solving problems, visualizing analytics, and modeling network features.

Workflow

1. Explore the ArcGIS Pro interface.
2. Explore ArcGIS Pro tabs.
3. Explore the ArcGIS Geoprocessing toolkit.

CHAPTER 2
Basic editing in ArcGIS Utility Network

Objectives

- Evaluate basic editing skills in ArcGIS Utility Network.
- Use the Validation tool to check for errors.
- Use new versioning tools specific to branch versioning.

Introduction

Utilities have a range of equipment and assets that serve a specialized purpose. ArcGIS Utility Network provides users with a unique insight into displaying and studying infrastructure in the GIS ecosystem. This chapter enables you to explore the operational behaviors of the network and utility assets that are modeled in the **Foundation Network** dataset. We will use editing tools to place simple features on a map and explore the basic components of an active utility network.

> **Tip:** For information about how to set up your utility network project, see the "Getting Started" section in the front of the book. This chapter uses the Water Distribution Utility Network Foundation dataset.

Tutorial 2-1: Walking through the basics of editing

1. From the **Datasets_For_UN_Skills_Book** folder, open **Water Distribution Utility Network Foundation.aprx**.

2. Open the **Water Distribution Editor** map.

Open the Create Features panel

3. On the ribbon, click the **Edit** tab.

4. In the **Features** group, click the **Create** button to open the **Create Features** pane.

   The **Create Features** pane enables users to preconfigure attributes prior to creating new features in the map space. Find **Templates** using the search dialog box and browsing through the list.

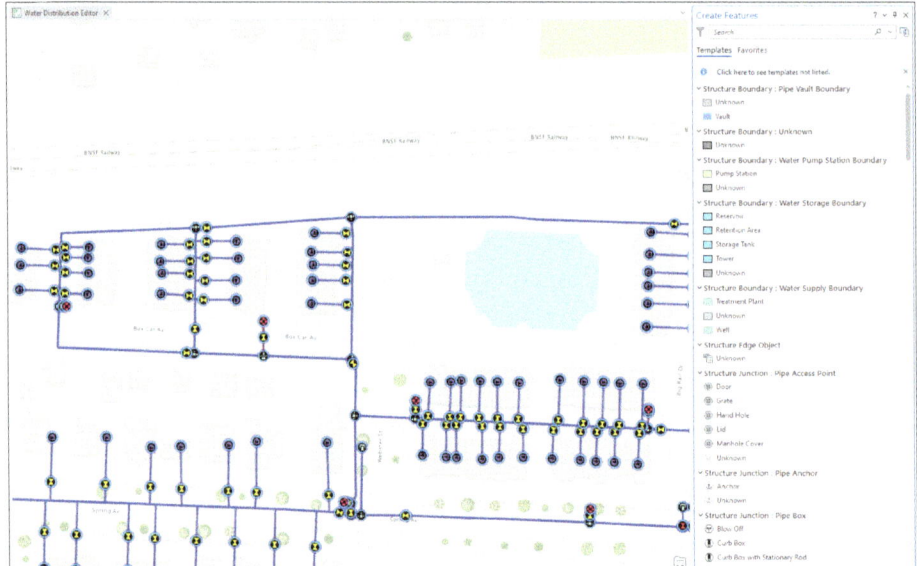

Update attributes in a feature

5. In the search dialog box of the **Create Features** pane, type Distribution Main and press **Enter**.

6. In the results, click **Distribution Main** to reveal its attributes and values.

   In this case, the value for the pipeline's **Diameter** is listed as <**Null**>.

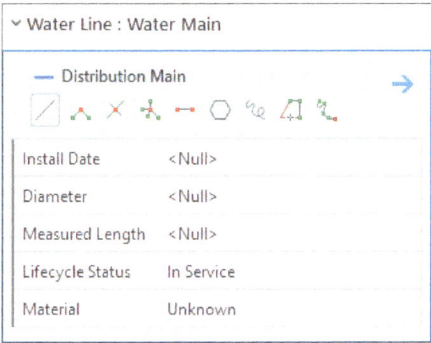

7. Double-click **Distribution Main** to open the **Active Template** pane.

   Working in the **Active Template** pane allows you to view the **Required** edits and opens the list of **Optional** edits for all attributes that can be edited.

8. Under the **Required** section, apply the following settings:

   - Set the **Install Date** to today's date.
   - Set the **Diameter** of the pipe to 4".
   - Change the **Material** to Galvanized Pipe – GP (or any material of your choice).

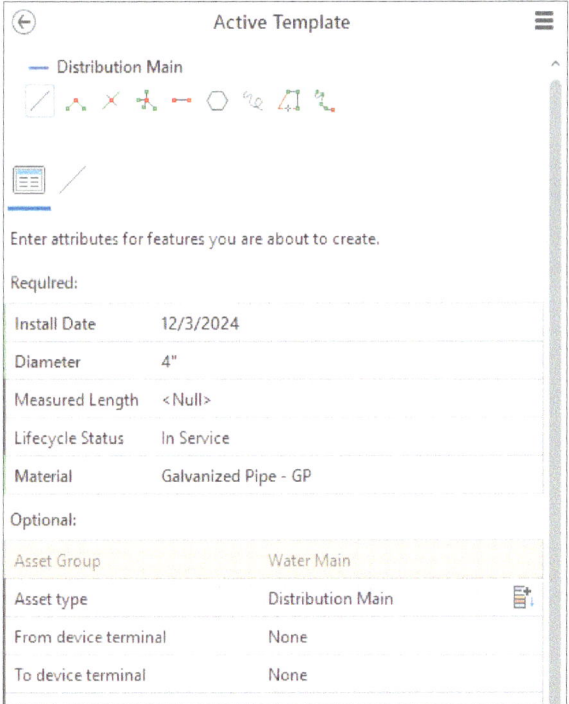

Adding a new line on the map

Now that all the attributes that you want to set for the new feature have been added, you'll add a new line on the map.

9. Under the **Distribution Main** name, click the **Line** tool.

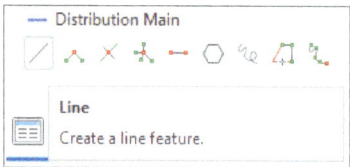

10. On the map, in the empty space above the **BNSF Railway**, click to add the first point of your line.

11. Create a line parallel to the **BNSF Railway**, adding a second point above the water feature.

12. Add a last, third point at an angle perpendicular to the line you just made.

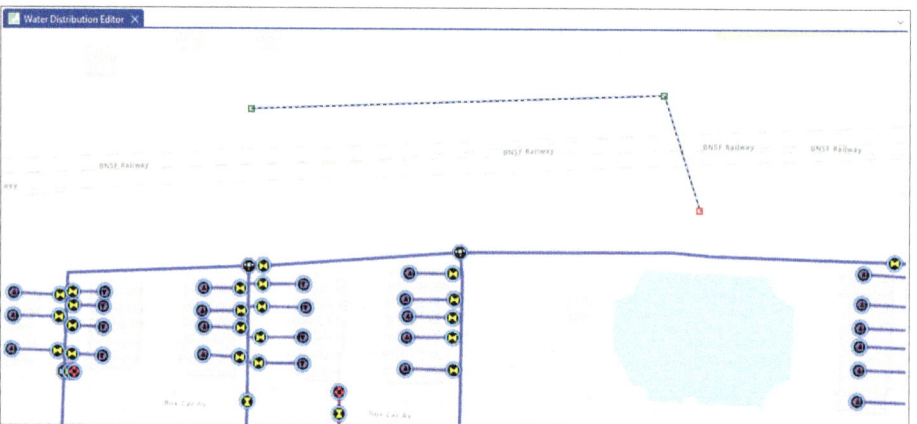

Introducing dirty areas and the network index

After drawing the new line, you may notice that the angle points (vertex) of the line remain highlighted and edited.

13. Right-click the last point you added (highlighted in red) and click **Finish**.

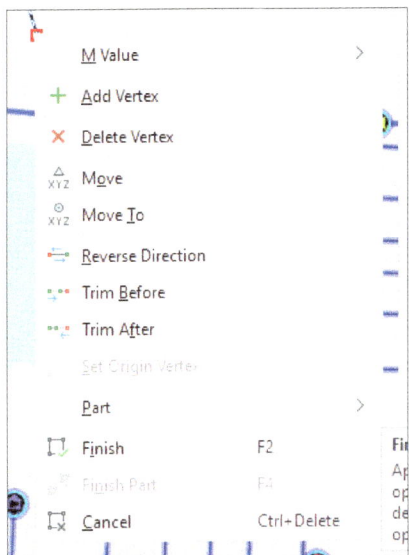

As soon as the line edit is completed, a purple hatched box or dirty area is generated on the map. These boxes represent map elements that haven't been validated and loaded into the network index. The process of validation checks to see whether the new features violate any existing network rules (see chapter 9) or improperly configured attributes for the new feature.

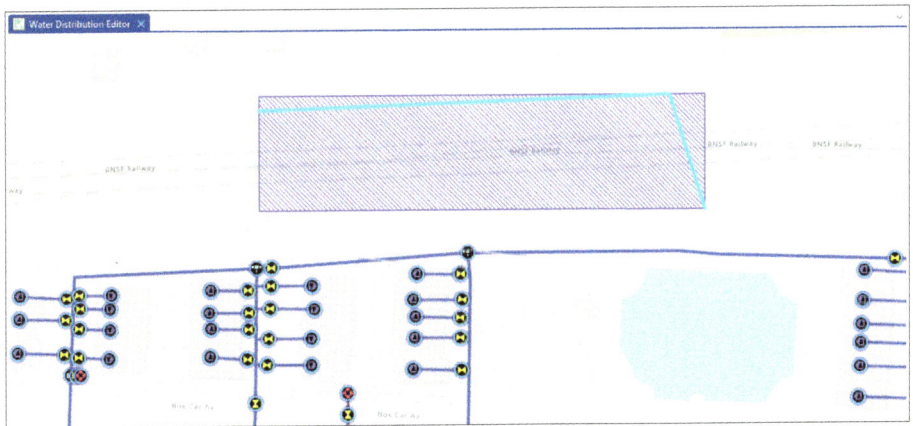

Completing feature validation

The network index enables users to return queries (via **Trace**) along with most of the other functions of ArcGIS Utility Network. To test the new edit and include the new line with the index, you will use the **Validate Network Topology** tool on the **Utility Network** tab.

14. On the ribbon, click the **Utility Network** tab.

15. In the **Network Topology** group, click the **Validate** button to run the **Validate Network Topology** tool.

On completing the validation process, the dirty area is removed, and the new lines can be traced.

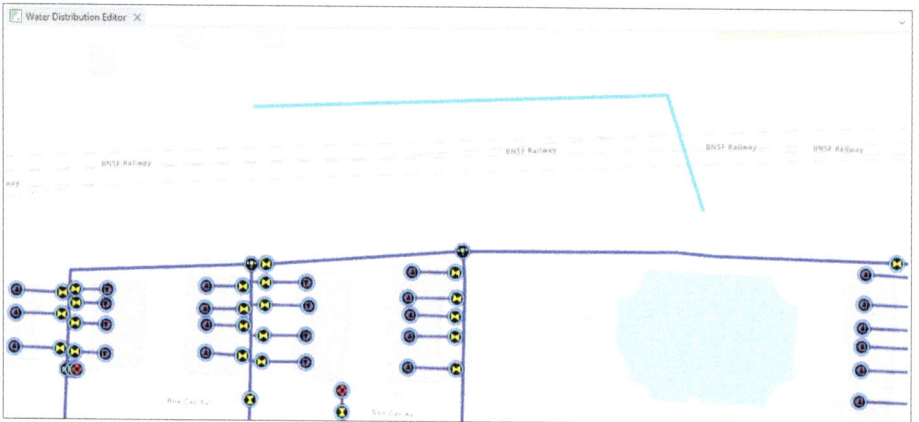

On your own

Create new lines and points to discover the editing process for the utility network.

Take the next step

Edits and validations are the core elements of ArcGIS Utility Network. Each of the **Foundation Network** datasets has diverse features, rules, and configurations that can expand your understanding of editing ArcGIS Utility Network features.

Summary

Network editing is a critical skill for ArcGIS Utility Network. It's important to recognize that map features aren't added to the network index until the data has been successfully validated. To improve your editing experience, it's best to preconfigure attributes prior to making the map edits.

Workflow

1. Open the Create Features pane.
2. Select a line feature and modify attributes.
3. Draw the line feature on the map.
4. Validate the dirty area to include the edit in the network index.

CHAPTER 3
Creating feature templates

Objectives

- Describe editing capabilities in ArcGIS Utility Network and explain the use of templates.
- Configure group templates for utility network features.
- Configure preset templates from selected features.

Introduction

ArcGIS Utility Network takes a unique approach to editing and managing data. As one of the first Esri products based on services architecture, it uses a fixed schema to model the range of network assets that utilities operate. In many cases, assets are built and deployed as an assembly of features. In lieu of building those complex features in every instance, group templates and preset templates from selected features are used to enable an optimal user experience. For example, a unit of equipment (such as a pole) might consist of dozens of individual features (spatial and nonspatial). Instead of requiring a drafter or GIS technician to place each feature, the features can be added to a map space as a cluster of related features. This chapter covers how to configure templates to model complex features.

Chapter 3: Creating feature templates

Tip: For information about how to set up your utility network project, see the "Getting Started" section in the front of the book. This chapter focuses on the Electric Utility Network Foundation dataset.

Tutorial 3-1: Building feature templates using group templates

1. From the **Datasets_For_UN_Skills_Book** folder, open **ElectricUtilityNetworkFoundation.aprx**.

2. From the **Catalog** pane, open the **Electric Network Editor** map.

Navigate to the Template tab

The **Template** tab is an element inside the **Create Features** pane. It allows templates to be established based on existing map layer options.

3. On the **Edit** tab, open the **Create Features** pane.

4. In the **Create Features** pane, to the right of the search box, click the **Manage Templates** button.

   The **Manage Templates** pane opens.

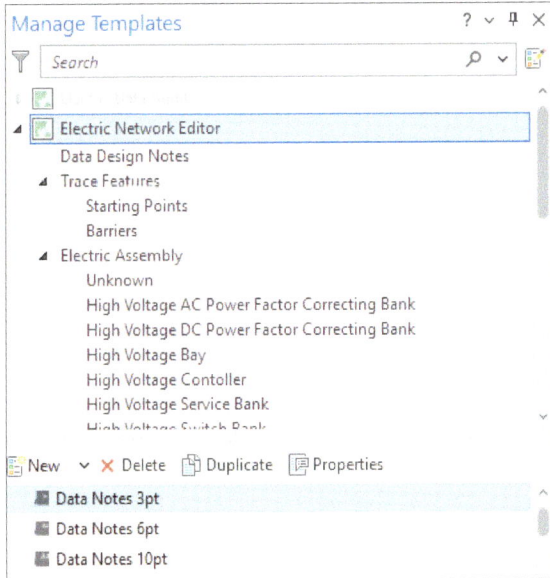

Explore the role of templates

In ArcGIS Pro, templates are used to enable users to preconfigure the default properties of features in the **Create Features** pane. In the simplest example, users can set the default attribute properties for a feature before creating the feature in the map space. For this skill, you'll use the **Manage Templates** pane to configure default attributes and cluster (or group) features into commonly used assemblies. In the utility network, assemblies are clusters of equipment that have a common purpose. With assemblies, you use an anchor feature as the foundation for the group of features.

To create a group template from the **Manage Templates** pane, you'll select the primary feature description (**Asset Group**) and identify the detailed feature description (**Asset Type**) from the bottom frame.

> **Tip:** For more details about how subtypes, asset groups, and asset types work, see chapter 5 ("Applying Feature Subtypes Using Asset Groups and Asset Types").

5. In the **Manage Templates** pane, expand the **Electric Network Editor** map folder. Locate and expand the **Electric Line** group. Select **Low Voltage Service**.

6. At the bottom of the pane, select **AC Overhead Service LV**.

   In many utilities, the most common data edit is the creation of a service drop for a home or a business. In this case, the **AC Overhead Service LV** represents a specific set of features that are preconfigured to connect a local residence to the adjacent distribution line.

7. Click the **New** button and select **Group Template**.

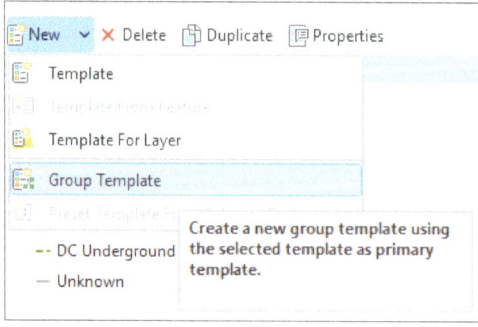

Create a group template for general

When using the **Group Template** tool, a **Template Properties** dialog box appears. In this dialog box, **Name** is the only required field. The **Description** and **Tags** fields can be used to locate the new template in the **Create Features** pane. Also, the interface lets you see the default symbol for the anchor feature.

8. In the **Template Properties** dialog box, for **Name**, type Residential Service Connection.

   This template will be used to enable GIS teams to create a residential service drop with fewer clicks. The critical features are already connected without manual intervention.

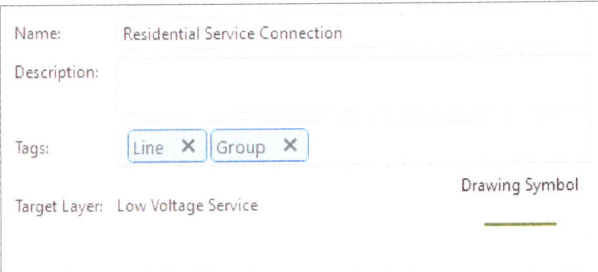

Create a group template for tools

The **Tools** tab establishes how the new grouped feature is placed using the **Create Feature** pane.

9. Click the **Tools** tab.

   For the **AC Overhead Service LV** feature, the default property is to place the set of features as a line on the map space. Other line drawing options are available as needed.

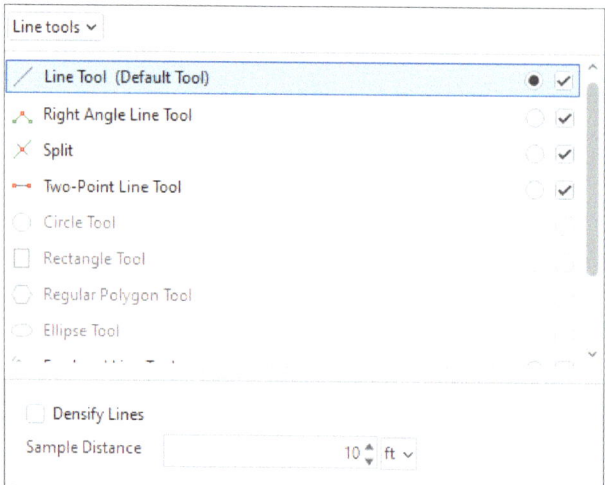

Create a group template for builders

The **Builders** tab is the most significant element when creating new group templates. It lets you define what additional features are included in the group and how those features behave when the edit process is run.

These builders set behavior (such as where and how features are connected in the template) that enable users to establish where and how the included template features are placed.

10. Click the **Builders** tab.

11. Click the **Add** button to see a list of available features to add to the group template.

    The service drop for a commercial site (restaurant, grocery store, or similar facility) is a little different from the service connection to your home. You'll open the **Commercial** example to explore the power of template builders.

12. Click **Choose a feature template** and search for Commercial. Select the first result.

Chapter 3: Creating feature templates

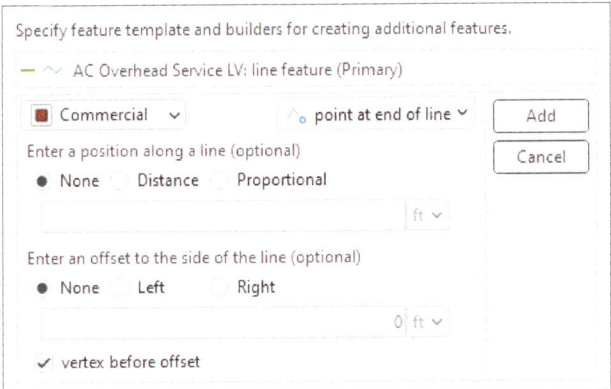

After you select the feature template, you may choose a builder to define the behavior or the placement of the feature on the group feature. In this case, you should place the customer service point at the end of the low-voltage service line. You have other options for placing the point at the beginning or midpoint of the line, vertex locations, or intersections. You can control the placement of the point by line or proportional distance from the builder spot or even offset the features as needed.

13. Click **point at the end of line** to open the builder list and review the different options.

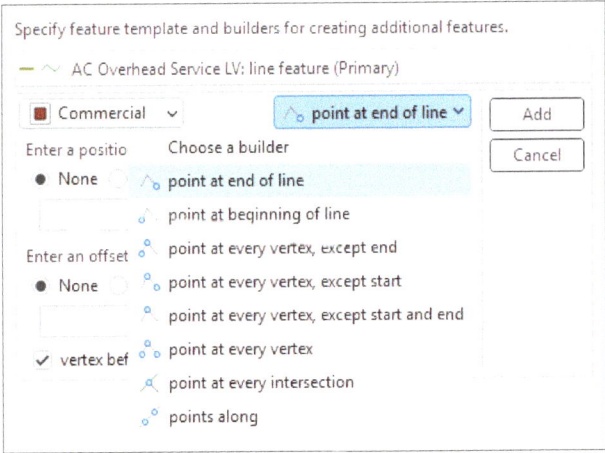

14. Select **point at end of line** as your builder.

15. Click **Add**.

You can add features and builders to each of your configured group templates.

This approach lets you build your assembly in simple or complex ways, as required by your workflow.

Create a group template for associations

The last element of the **Template Properties** window is the **Associations** tab. The **Add associations when creating new assets** check box enables the **builder** to create associations if the group includes features that may be contained as part of the builder process.

16. Click the **Associations** tab.

17. Ensure that the box for **Add associations when creating new assets** is checked.

   > Primary template: Low Voltage Service : AC Overhead Service LV
   > ✓ Add associations when creating new assets

18. Click **OK** to create your new group template.

   Creating group templates for utility network features enables you to create sophisticated features using preconfigured templates. For the most part, this approach is best used for building elements along linear features. For point features, the option to use a preset template from selected features may be more suitable.

Tutorial 3-2: Building feature templates using preset templates from selected features

Create features to be selected

The option to use a preset template from selected features differs from the group template approach in that you build the features for the template in advance. In this case, the elements of a pole top transformer bank are built to include fuses on the high side, with the individual transformers for each phase and the pole for the elements attached.

1. In the **Electric Network Editor** map, zoom to an empty, blank area that has no features.

2. Open the **Create Features** pane. Search for the Vault with 4 walls feature template.

To understand the current setting of the template, you'll drop an example of this template on the map.

> **Tip:** Vaults and boxes for electric systems are typically small objects (spatially speaking). To visualize the details of this template, a large scale (1:5) is used to visualize the features on this map.

3. On the map, with the **Point** tool selected, double-click to place a vault.

   A dirty area will appear around the vault you just placed.

4. On the **Edit** tab, in the **Manage Edits** group, click **Save**.

5. On the **Utility Network** tab, in the **Network Topology** group, click **Validate** to run the **Validate Network Topology** tool.

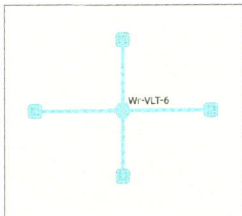

> **Tip:** The name assigned to your vault (for example, Wr-VLT-5) may differ from the one in the image.

For this tutorial, you will convert the **Vault with 4 walls** feature to a **Vault with 2 walls** feature by using the **Edit Vertices** tool and relocating two of the knockout points (square features) to the left side of the vault.

6. On the **Edit** tab, in the **Tools** group, click the **Edit Vertices** tool.

7. On the map, select the top line. Hover your pointer over the middle of the line, right-click the line, and click **Add Vertex**.

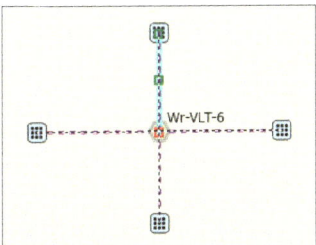

8. Click the knockout point at the end of the line and drag it to the left. Align the point so that the line is parallel to the one below it.

9. Repeat steps 7 and 8 for the bottom line.

10. Save your edits and run the **Validate Network Topology** tool. Refer to the following image for the result.

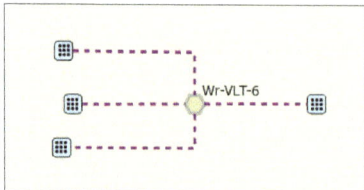

11. On the map, select all the features of **Vault with 2 walls** to highlight it.

Configure a new template for preselected features

Creating a new template for preselected features looks like the group template configuration.

12. Open the **Manage Templates** pane. Search for Wire Vault and in the results, click **Wire Vault**.

13. At the bottom of the pane, select the **Vault with 4 walls** feature template.

14. Click the **New** button and select **Preset Template From Selected Features**.

Create a preset template for general

On the **General** tab, the **Name** field is required.

15. For **Name**, type Vault with 2 walls.

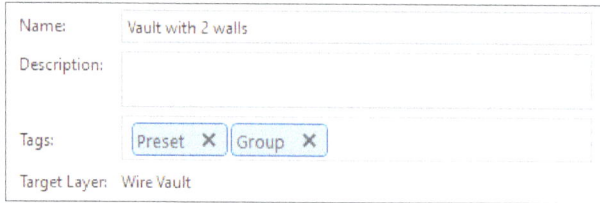

Create a preset template for tools

The **Tools** tab provides options and configurations for users when the new template is created.

16. Click the **Tools** tab.

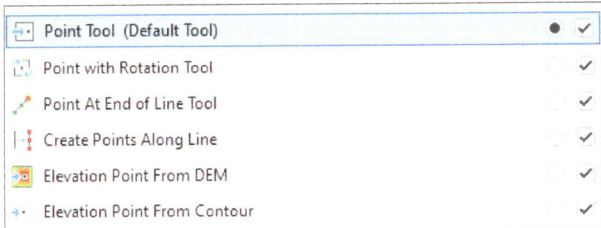

Create a preset template for features

The **Features** tab enables you to review all the selected features in the template and review or adjust attributes as needed.

17. Click the **Features** tab to look through the features and their associated attributes for the **Vault with 2 walls**.

Create a preset template for preview

Finally, you can preview the template before saving the **Preselected Feature Template** on the **Preview** tab.

18. Click the **Preview** tab to see what the template will look like.

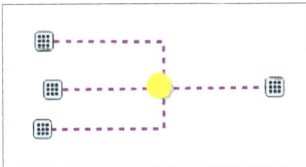

19. At the bottom of the window, click **OK** to create the new preset template.

On your own

Build a group template to include poles offset from line vertex locations for an overhead line.

Take the next step

Build a series of group templates and preset templates from selected features to model different assemblies in the **Gas and Pipeline** or **Water Distribution Network Foundation** dataset project.

Summary

Group templates and preset templates from selected features enable utility network users to build complex assemblies based on commonly used asset configurations. Both editing configurations were used to enable users to have an optimal user experience, better performance, and higher-resolution data as needed to support their workflows.

Workflow

1. Identify a line option for a group template.
2. Use builders to place points at vertex points and offset as needed.
3. Create an assembly of lines and devices to select for a preset template.
4. Identify an assembly feature to build the preset template.
5. Add associations in the template configuration.

CHAPTER 4
Overview of structure and domain networks with tiers

Objectives

- Describe the difference between structure and domain networks.
- Demonstrate how tiers reflect the unique properties and restrictions of a business unit.
- Review partitioned and hierarchical tiers and explore the difference in configurations.

Introduction

One of the key elements of utility networks is the way the data is modeled and configured. This chapter is intended to walk through the various schema components and demonstrate how they are applied to model utility assets. When utilities represent assets in a GIS (in a schema), many times they consist of features that have operational significance (provide flow, or product). In other cases, features support the operational elements. All those features are represented in a configured schema for performance.

> **Tip:** For information about how to set up your utility network project, see the "Getting Started" section. This chapter features the Gas and Pipeline Utility Network Foundation dataset.

Tutorial 4-1: Exploring the structure network

1. From the **Datasets_For_UN_Skills_Book** folder, open **Gas and Pipeline Utility Network.aprx**.

2. From the **Catalog** pane, double-click the **Gas and Pipeline Network Editor** map to open it.

   Utility assets often have features or components that support or have an impact on reliability. They can be key elements that are inspected and maintained the way their counterparts are operated to provide service to customers. In the utility network, these features are modeled in structure networks that exist in every utility network. The structure network consists of one point layer (structure junction), one line layer (structure line), and one polygon layer (structure boundary), as well as two tables to support points and lines that share geometries (structure junction object, structure edge object).

3. In the **Contents** pane, expand the **Structure Junction**, **Structure Line**, and **Structure Boundary** layers.

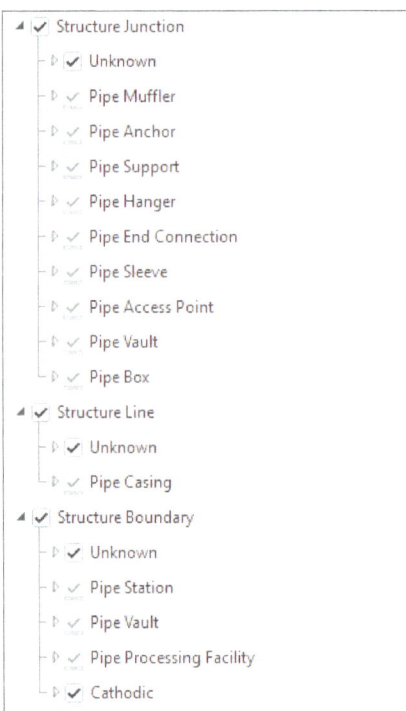

Explore structure junction features

4. Within the **Structure Junction** layer, right-click the **Pipe Box** layer and click **Attribute Table**.

   This will return one record in the table.

5. Select the record and press **Ctrl+I** to open the **Pop-up** information window.

6. At the top of the **Pop-up** window, click the arrow next to **Pipe Box** to expand the list of objects.

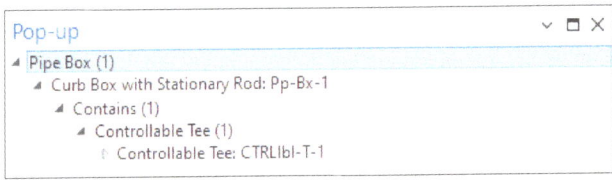

   The **Pipe Box** has internal objects. Other examples of junctions might include anchors, braces, or other supporting devices.

7. Close the **Pop-up** window and the attribute table.

Explore structure line features

The **Gas and Pipeline Network Foundation** map consists of a gathering zone to show how wellheads can be modeled and connected to transmission pipeline features. One example of a structure line feature is the pipe casing feature in the **Structure Line** layer.

8. On the ribbon, click the **Map** tab. In the **Navigate** group, click **Bookmarks** and select the **Gathering Well** bookmark.

   The map will zoom to the **Gathering Well** location.

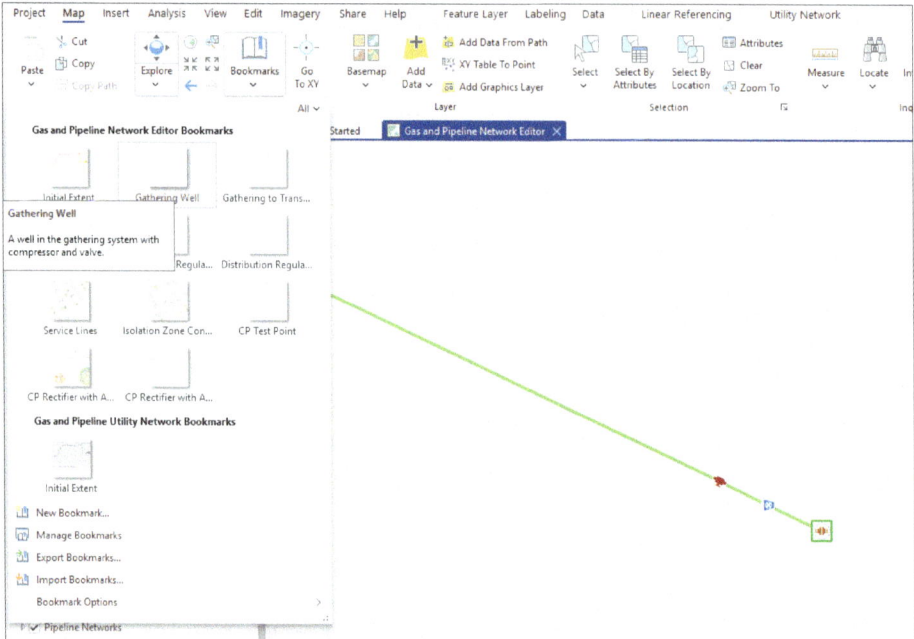

Open the Create Features pane

Pipe casing is used to provide structural integrity for the gas to flow to the collection elements of the wellhead. You will create a casing feature to be represented on the map.

9. Click the **Edit** tab. Within the **Features** group, click **Create** to open the **Create Features** pane.

10. In the **Create Features** pane, search for Casing to return the **Structure Line : Pipe Casing** feature template.

11. Click the **Casing** feature.

12. Beneath the feature name, select the **Line** tool to create a line from the wellhead in the center of the **Map** view in any direction.

Chapter 4: Overview of structure and domain networks with tiers

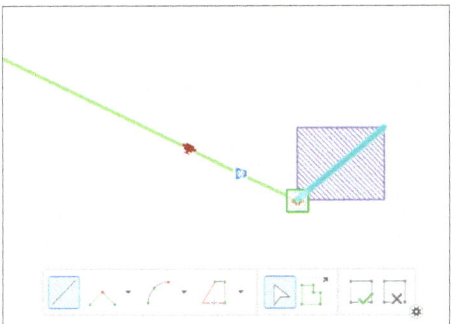

13. At the bottom of the map, on the **Configure** toolbar, click the **Finish** button (a square with a checkmark).

Explore structure boundary features

The **Gas and Pipeline Network Foundation** has a transmission compression station where the gathering zone facilities are shifted to transmission pipes. The **Structure Boundary** layer has a plant station to represent the station as a facility.

14. From the **Bookmarks** list, select the **Gathering to Transmission Compression Station** bookmark.

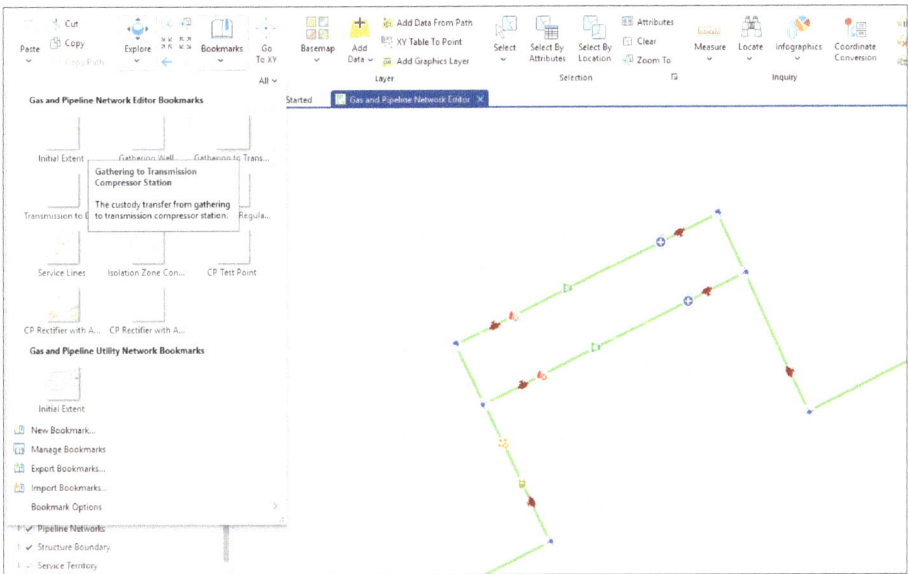

15. Open the **Create Features** pane.

16. Using the search bar, type Station Structure to return the **Structure Boundary : Pipe Station** feature class. Click the **Station Structure** feature.

17. For **Station Name**, click **<Null>** and type Baseline Station as the name of the facility. Press **Enter**.

18. Under the feature name, click the **Polygon** tool.

19. On the map, use the **Polygon** tool to create a box around all the features of the facility.

    This box will become the pipe station for the transmission compressor station.

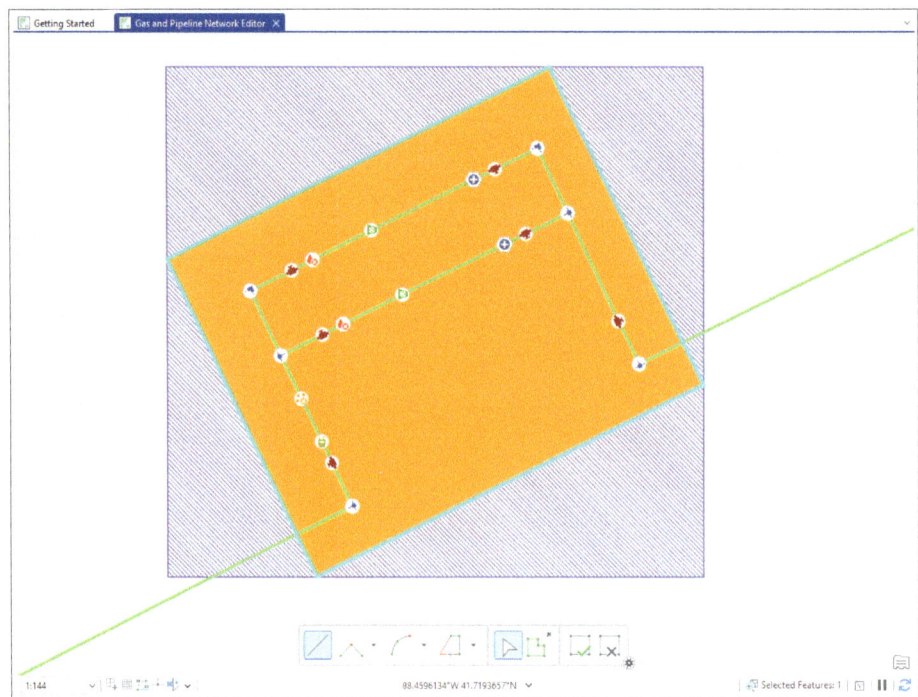

20. Click the **Finish** button.

21. On the **Edit** tab, in the **Manage Edits** group, click **Save** to save your edits.

Tutorial 4-2: Working with domain networks

The utility network is often described by a commodity or domain, such as electric, gas, water, or communications. The behavior of electric power down a line is significantly different from gas or water pipes. Domain networks have a number of feature layers to model the unique properties of utility assets that have operational properties.

Explore pipeline device features

The **Device** layer is unique in that features modeled as devices have action, can enable or disrupt flow, and can meter, monitor, or modify flow of electricity, gas, water, or any commodity that's modeled. One example of a pipeline device can be found by using the **Distribution Regulator Station** bookmark.

1. Open the **Bookmarks** list and click the **Distribution Regulator Station** bookmark.

2. On the map, toward the left side of the station, click the **Relief Valve** (orange valve symbol) to view the **Pop-up** window.

   You can tell that this is a pipeline device by looking at the **Device Status** field. It shows as **Open**, which means that it's flowing. The **Relief Valve** can be converted to a closed state or bypassed using other valve configurations.

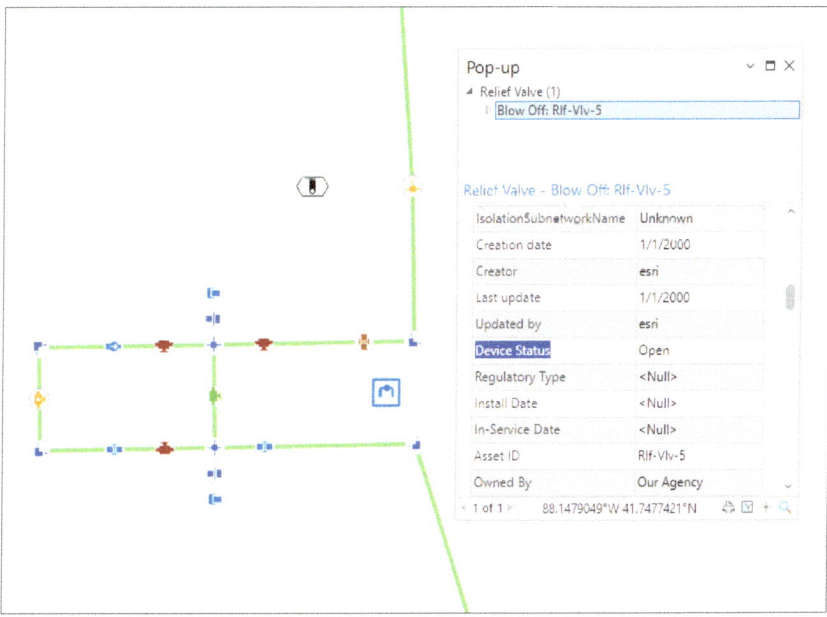

Explore pipeline junction features

The **Junction** layer represents points that are like devices. They are connected features but don't have action. They are akin to connectors, sleeves, flanges, and elbows. You can explore pipeline junctions using the same **Distribution Regulator Station** bookmark.

3. In the **Contents** pane, expand the **Pipeline Junction** layer. Right-click the **Elbow** layer and click **Selection** > **Select All**.

   All the elbows in the station are now highlighted.

4. Click one of the selected elbows and scroll through the attributes in the **Pop-up** window.

   You can see that these devices are in a fixed-flow configuration.

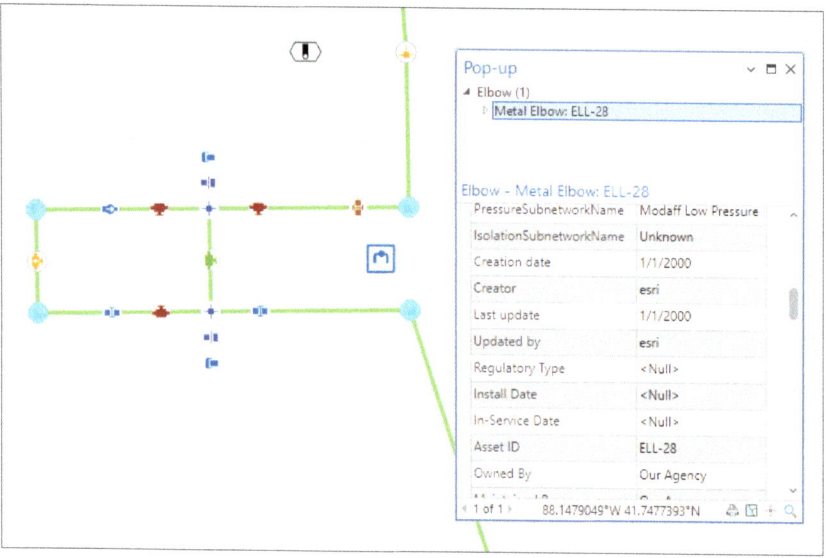

5. Click the **Map** tab. In the **Selection** group, click **Clear** to clear your selection.

Explore pipeline assembly features

Assemblies are point features that function as containers, such as vaults, bays, or housings for devices and junctions that have a common purpose. Using the same **Distribution Regulator Station** bookmark that you used for the devices and junctions, you can identify the station's assembly feature.

6. On the map, toward the right of the station, click the **Assembly** feature (square symbol).

7. In the **Pop-up** window, review all the features that are contained by the facility.

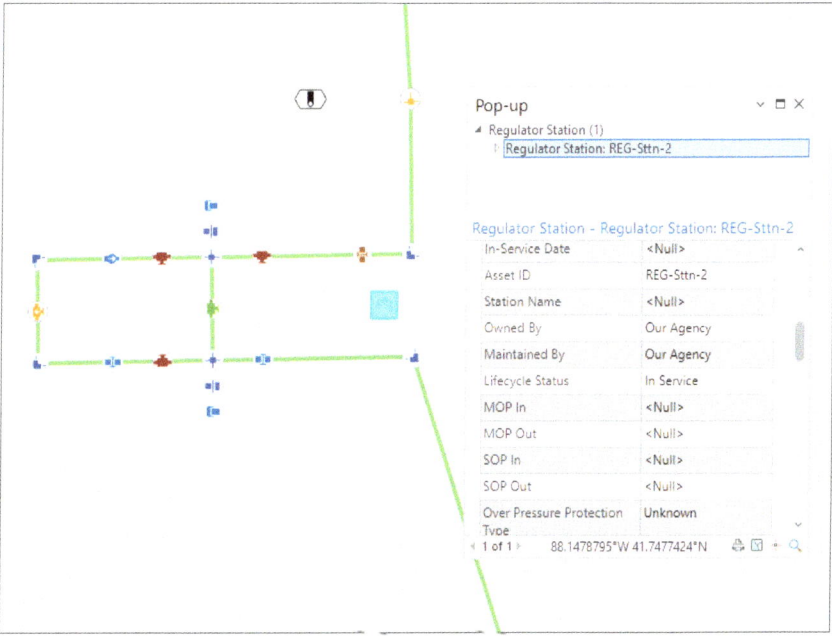

Explore pipeline line features

The **Line** layer represents pipelines, cables, or wires that are the primary conduit for the commodity specific to the utility model. **Line** features occur at all scales and capabilities for any utility model.

8. On the map, click any line feature and review its **Pop-up** window.

   In the **Distribution Regulator Station**, the line features represent station pipes.

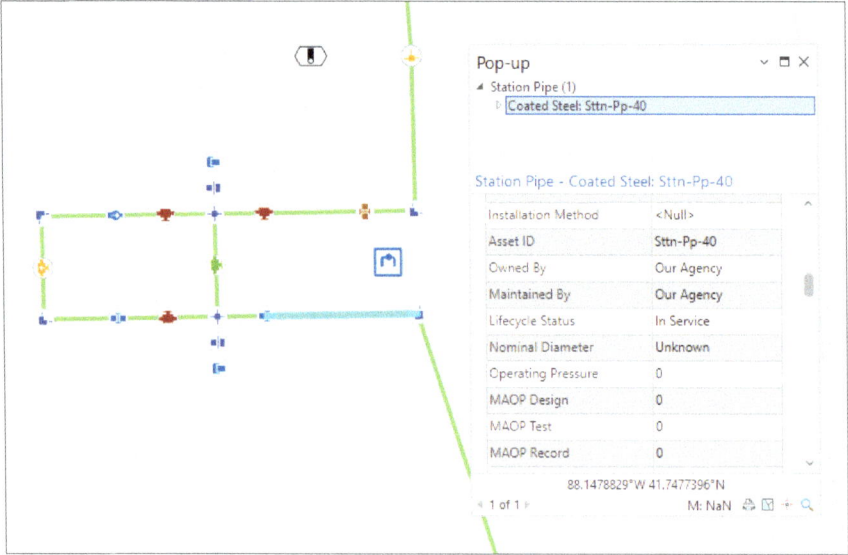

Explore pipeline subnetwork lines

Subnetwork lines represent the feeders or circuits and are a repository of key attributes from all the domain features (**Devices**, **Junctions**, **Assemblies**, and **Lines**). To see an example of a subnetwork line, you'll browse the **Washington Pressure** subnetwork.

9. In the **Bookmarks** list, click **Washington Pressure Subnetwork**.

10. Click the **Utility Network** tab. In the **Subnetwork** group, click the **Find** button to open the **Find Subnetworks** pane.

11. In the list of subnetworks, find the **Washington Pressure** subnetwork.

12. Right-click the subnetwork and click **Trace Subnetwork**.

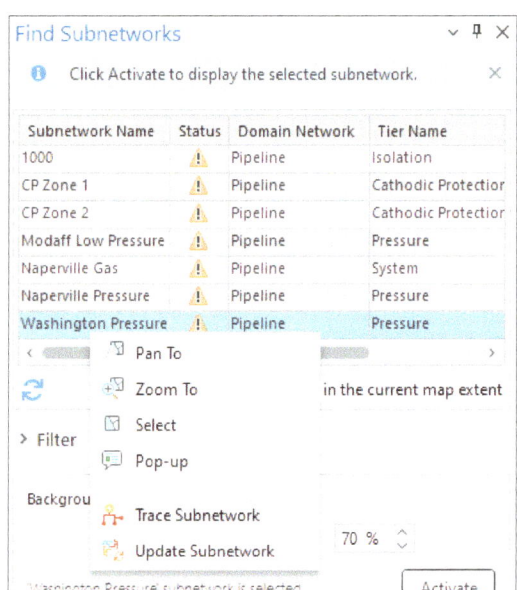

13. Right-click the subnetwork again and choose **Pop-up** to open the **Pop-up** window.

Explore nonspatial objects

Tables are provided in the utility network to enable objects that augment the structure network and any type of network domain. These tables can fully participate with structure and domain network features. For each domain, there's a single edge object (representing lines with shared geometry) table that models lines using nonspatial records and a single junction object (representing point features with shared geometry) table that can be used to model devices, assemblies, or junctions as nonspatial records. Modeling features using nonspatial objects can be highly performant when working with large datasets and when representing features with a shared geometry or congested map area. For examples of nonspatial objects, consider fiber strands in a cable (domain network) or ducts in a duct bank (structure network).

Tutorial 4-3: Examining tier definitions

Tiers are used to organize and manage data in domain networks by defining a shared set of properties and restrictions for a set of subnetworks. Two types of tier definitions are available: partitioned and hierarchical. The type of product that's served by the domain network (for example, gas, electric, or water) helps guide your decision on whether to use a partitioned or a hierarchical tier definition to model your subnetworks. Both tier definitions establish the type of assets represented and the behavior and connective properties of those elements in the tier.

Explore partitioned networks

Partitioned network definitions and support systems represent wire systems that are sequential and often mutually exclusive, such as electric or communications. For example, users may manage an electric domain network using multiple tiers for different voltages ranges. They may have a high-voltage tier with lines ranging from 500 kV to 46 kV, a medium voltage tier ranging from 38 kV to 4 kV, and a low voltage tier ranging from 2.5 kV to 230 V. For each of these tiers, the assets can exist only in a single tier and cannot be used across multiple voltage ranges at the same time. As a result, each feature is modeled separately.

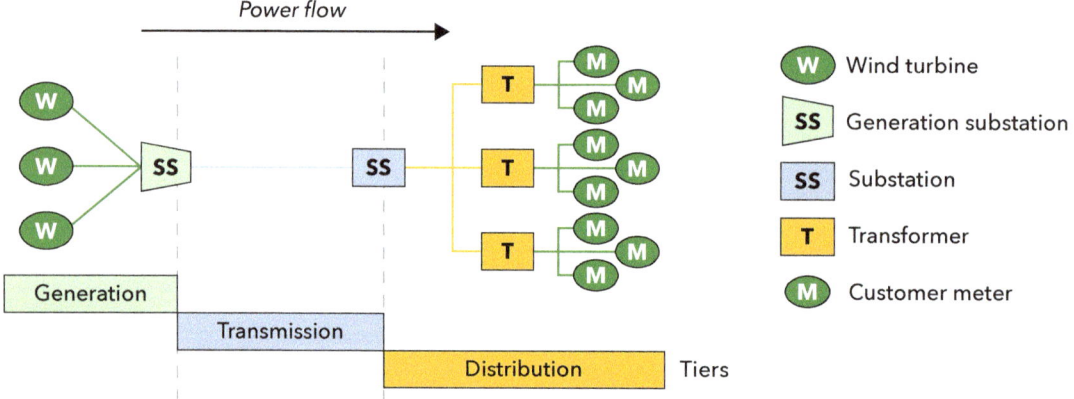

Explore hierarchical networks

Hierarchical network definitions are usually used to represent pipe systems, such as gas or water networks. A domain network with a hierarchical tier definition may contain nested tiers (where it makes sense to visualize different operational states). For example, an isolation tier may be nested inside a pressure tier, and both of them may be nested within a cathodic protection tier.

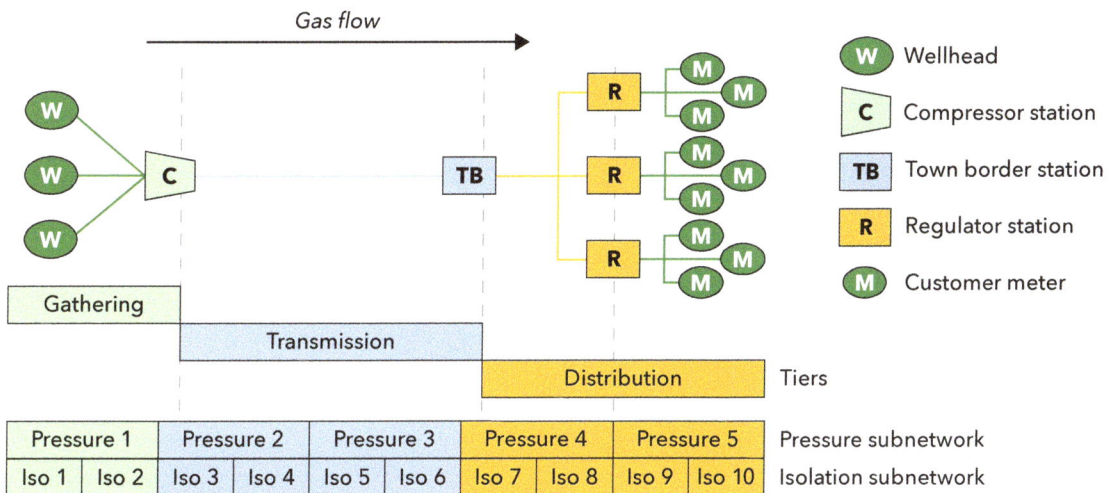

On your own

Explore the layers, tiers, and domain configurations for the other network foundations.

Take the next step

Try editing features and changing attributes to see how they impact and interact with each other.

Summary

In this chapter, you examined the schema and data format of the utility network.

Workflow

1. Explore the structure network.
2. Work with the domain network.
3. Describe nonspatial objects.
4. Describe domain configurations.

CHAPTER 5
Applying feature subtypes using asset groups and asset types

Objectives

- Describe how the utility network data model can be configured to represent all assets and features in a fixed schema.
- Configure layers to increase performance for your network model.
- Apply asset groups and asset types across operational and structure domains.
- Use feature templates.

Introduction

ArcGIS Utility Network uses a fixed schema for the classes that are created in a domain or structure network (see chapter 4 for more details). This schema is a significant departure from traditional GIS, in which an individual feature class is used to represent a single asset configuration. With only five feature classes to represent operational assets in the domain network (**Devices**, **Junctions**, **Assemblies**, **Lines**, and **Subnetwork lines**), a new approach is needed to represent the various types of assets and their configurations because they may be used in the scope of a GIS (2D, 3D, or digital twin).

Tip: For information about how to set up your utility network project, see the "Getting Started" section. This chapter features the Gas and Pipeline Utility Network Foundation and the Electric Utility Network Foundation datasets.

Tutorial 5-1: Introduction to subtypes

ArcGIS Utility Network uses subtypes to represent asset groups. A subtype represents a collection of features that share attributes. In the context of ArcGIS Utility Network, this functionality serves the same purpose and represents the major classification of assets.

Open the database to explore the layers

1. From the **Datasets_For_UN_Skills_Book** folder, open **Gas and Pipeline Utility Network Foundation.aprx**.

2. In the **Gas and Pipeline Network** editor map, browse to the **Catalog** pane.

3. Expand the **Databases** folder.

4. Expand the **UPDM_UtilityNetwork.gdb** file geodatabase, and expand the **UtilityNetwork** feature dataset to locate the **PipelineDevice** feature class.

Explore the data design tools

5. Right-click the **PipelineDevice** feature class and click **Data Design**.

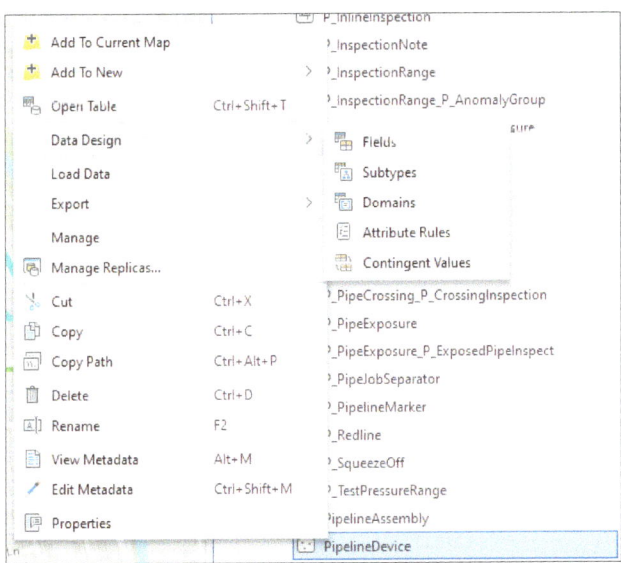

This context menu provides several options to view and configure schema and define how this class is modeled. You can create or configure fields, subtypes, and domains; author attribute rules; or define contingent values for your data.

6. Click each type of data design to view how the data can be edited.

Explore the subtype tools

The **Subtypes** view for the **PipelineDevice** class provides the feature-class-level subtypes (asset groups) arrayed in column form with the ability to set attribute domains and default values for each.

7. From the **Data Design** context menu, click **Subtypes**.

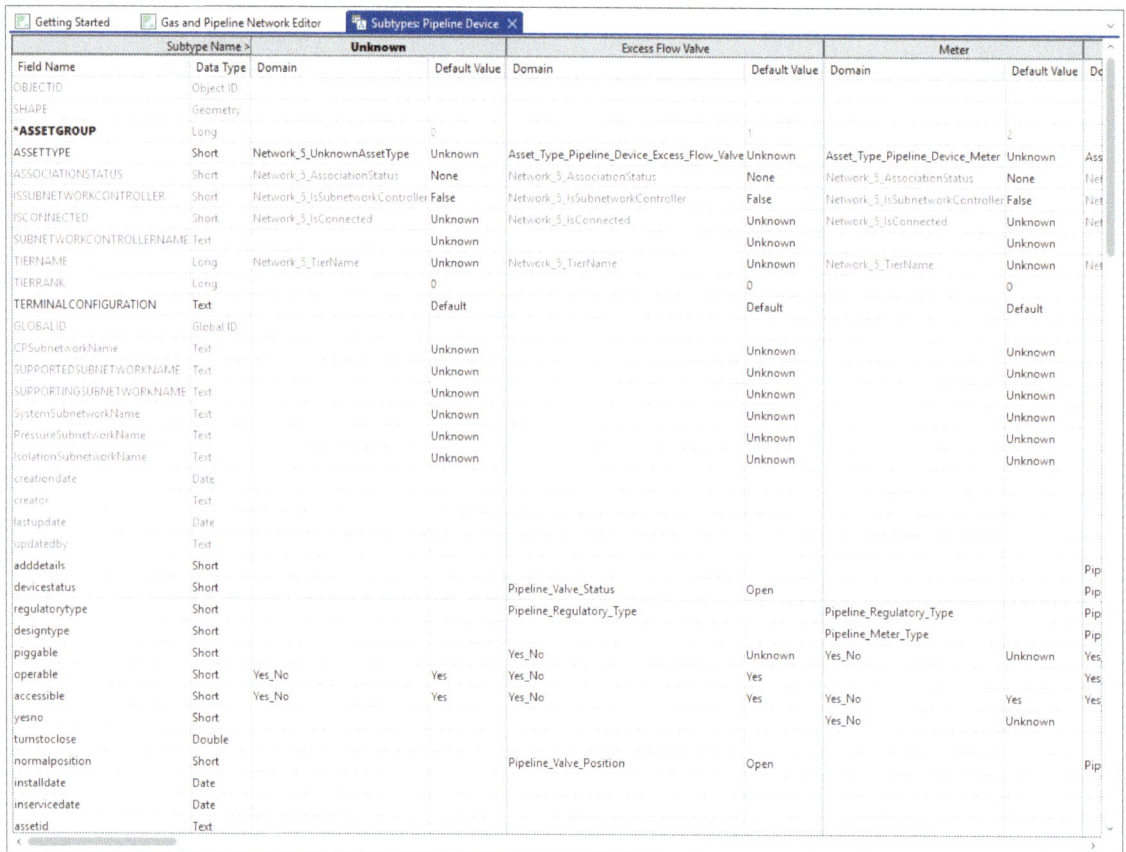

Manage subtypes

The **Manage Subtypes** window provides the ability to set the general parameters for the asset group.

8. On the ribbon, on the **Subtypes** tab, click the **Create/Manage** tool to open the **Manage Subtypes** window.

   You can see the window displaying the various subtypes defined for the **PipelineDevice** feature class.

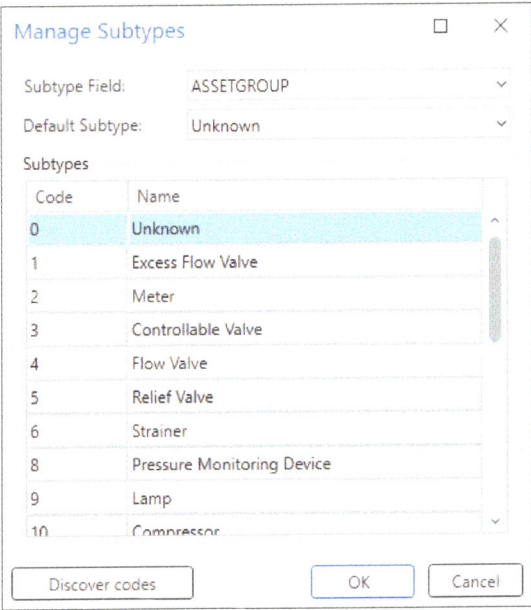

9. Close the tool.

Configure asset types for an asset group

Whereas an asset group is the subtype of the feature class, asset types are simply a domain-driven attribute exposed within each subtype. The **Pipeline Line** layer is a simple example of how layers are broken into an asset group (**Service Pipe**) and asset types (**Bare Steel**, **Cast Iron**, and so on). This approach gives utilities the ability to represent detailed and accurate map data in a more performative format under the **Pipeline Line** layer.

10. In the **Contents** pane, expand the **Pipeline Line** layer and the **Service Pipe** layer to review the different asset types.

Tutorial 5-2: Configuring default asset types

If asset groups provide a general description, asset types enable utilities to visualize specific data properties for the various lines, devices, and other elements.

1. In the **Catalog** pane, expand the **Databases** folder.

2. Expand the **Electric_UtilityNetwork.gdb** file geodatabase and the **UtilityNetwork** feature dataset to locate the **ElectricDevice** feature class.

Configure feature domains

3. Right-click the **ElectricDevice** feature class and click **Data Design**. Select **Domains** to open the **Domains** view.

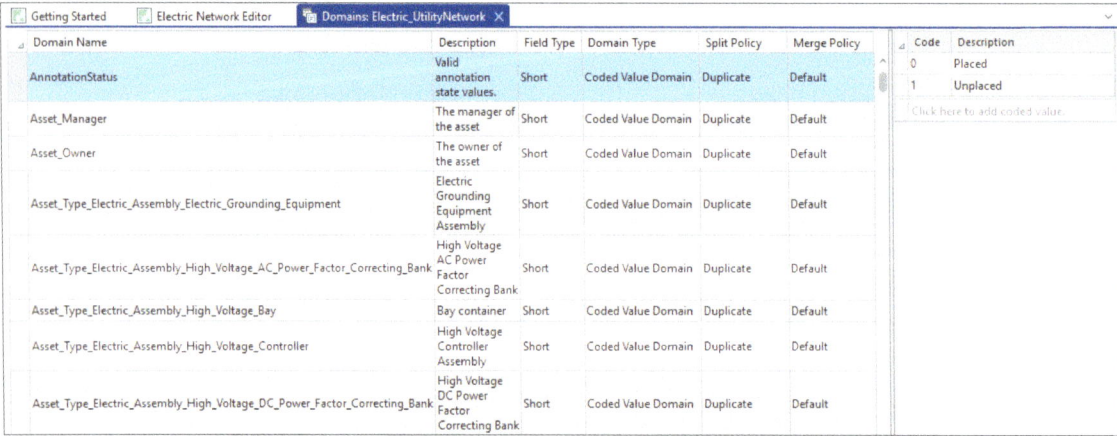

Explore domain values for an asset type

4. With the **Domains** view open, scroll down to locate the group of **Electric Device** rows.

5. From the **Domain Name**, find **Asset_Type_High_Voltage_Control_Unit**.

   This is the domain for the **High Voltage Control Unit** asset group. Looking at **Asset_Type_High_Voltage_Control_Unit**, the description represents the asset type values. It's important to note that the field type used for asset type domain-coded values is always the **Short** integer.

Asset_Type_Electric_Assembly_Medium_Voltage_Switch_Bank	Medium Voltage Switch Bank	Short	Coded Value Domain	Duplicate	Default
Asset_Type_Electric_Assembly_Medium_Voltage_Transformer_Bank	Medium Voltage Transformer Bank	Short	Coded Value Domain	Duplicate	Default
Asset_Type_Electric_Device_Ground	Ground	Short	Coded Value Domain	Duplicate	Default
Asset_Type_Electric_Device_High_Voltage_Control_Unit	High Voltage Controller	Short	Coded Value Domain	Duplicate	Default
Asset_Type_Electric_Device_High_Voltage_Fuse	High Voltage Fuse	Short	Coded Value Domain	Duplicate	Default
Asset_Type_Electric_Device_High_Voltage_Generation	High Voltage Generation	Short	Coded Value Domain	Duplicate	Default

Assign domain to asset type in the Subtypes view

Now that the asset types have been defined in the **Domains** view, you can see how those properties are set for each asset group in the **Subtypes** view.

6. In the **Catalog** pane, right-click the **ElectricDevice** feature class and click **Data Design > Subtypes**.

7. At the bottom of the **Subtypes** view, scroll to the right so that the **High Voltage Control Unit** asset group is the first group.

   For the **High Voltage Control Unit** asset group, the **Asset_Type_Electric_Device_High_Voltage_Control_Unit** domain is assigned to the **ASSETTYPE** field with **Control Unit** set as for **Default Value**. The **Subtypes** view also enables the default domain values for the common attributes assigned to the asset group.

Field Name	Data Type	Domain	Default Value
OBJECTID	Object ID		
SHAPE	Geometry		
*ASSETGROUP	Long		3
ASSETTYPE	Short	Asset_Type_Electric_Device_High_Voltage_Control_Unit	Control Unit

On your own

While you are in the **Electric Device** layer, set default asset types for other subtypes (asset groups).

Take the next step

Explore asset groups and asset types in the other **Foundation** data types.

Summary

The major and minor classification attributes of asset group and asset type enable utilities to model all their assets using a small, fixed number of classes. Asset groups use subtypes to cluster features that share a set of attributes, whereas asset types allow you to further define utility assets with a high level of detail through the configuration of coded-value domains and default values.

Workflow

1. Browse to a layer to explore subtypes.
2. Explore asset group configurations.
3. Set default asset types for asset groups.

CHAPTER 6
Terminals and controllers

Objectives

- Configure terminals on devices.
- Set subnetwork controllers on a designated terminal.
- Demonstrate how terminals are logical elements assigned to devices.
- Demonstrate how network rules work by connecting a line to a terminal.
- Trace through a device with terminals to show directionality.

Introduction

Earlier in the book, devices were defined as assets in a domain network that have an impact on the resource passing through them. They can enable or disrupt flow, meter or modify commodity, and impact the direction of flow. Devices in the field define locations or connection points from which resources, such as electricity and water, flow in and out. A utility network allows you to model these connection points with terminals.

Every device in the field can be physically connected to one or more lines through network rules that establish what may connect to its terminals (in the case of edge-junction rules) or how connectivity is established between devices (junction-junction rules).

Subnetwork controllers are devices that are used to define the origin or termination of a subnetwork. They are designated as the source of a feeder or circuit (in systems such as gas or electric) or the sink (in gravity-fed systems such as storm or wastewater). For example, in an electric system, a subnetwork controller could be a power-generating station or substation. For a gravity-fed wastewater system, a subnetwork controller could be a pump station or treatment plant. A subnetwork controller is established on a terminal of a specific device whose asset types have been configured to act as a controller.

Tip: For information about how to set up your utility network project, see the "Getting Started" section. This chapter features the Water Distribution Utility Network Foundation dataset.

Tutorial 6-1: Configuring and setting terminals on a device

Terminals are the logical elements that establish direction across and through a device. For example, a pump (as a device) can have an input terminal and an output terminal that define what can get connected and how it flows though the device.

Explore terminals

1. From the **Datasets_For_UN_Skills_Book** folder, open **Water Distribution Utility Network Foundation.aprx**.

2. In the **Water Distribution Editor** map, browse to the **Map** tab.

3. In the **Inquiry** group, click the **Locate** button to open the **Locate** pane.

4. In the **Locate** pane, click the **Layer Search** tab and search for a water device with an **asset ID** of Prssr-Vlv-1 to find **Pressure Valve-Pressure Reducing: Prssr-Vlv-1**.

Chapter 6: Terminals and controllers   47

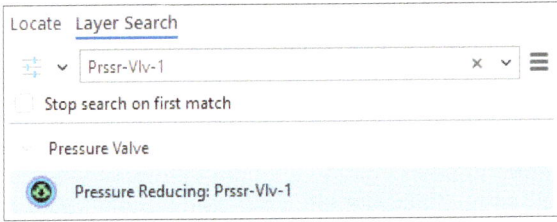

5. In the results, right-click the **Prssr-Vlv-1** device and click **Zoom To**.

6. In the bottom-left corner of the map, adjust the **Scale** to 1:75.

7. On the **Utility Network tab**, in the **Network Topology** group, click the **Terminal Paths** button to open the **Modify Terminal Paths** pane.

   This pane allows you to set the appropriate path from the assigned terminal configuration that controls how a resource may flow through the given device.

8. On the map, select the water device (black circle with green valve icon).

Modify terminal path options

With the **Modify Terminal Paths** pane open, you can see that the terminal configuration assigned to this device is called **Pressure Reducing**. The default path (A11) enables connections to be made through any terminal without restriction and facilitates flow in every direction through a given device. The **Path** drop-down list provides all the available valid paths that have been configured for use with the selected device. In this case, the other option is to block all paths (**None**).

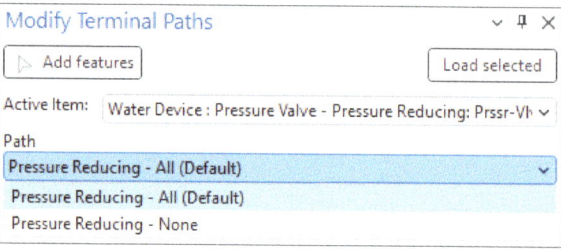

Configure terminal connections

A second tool for working with terminals can be found in the **Modify Terminal Connections** pane, which enables you to establish connectivity between a selected line and the appropriate terminals for the devices found at its endpoints.

9. On the **Utility Network** tab, in the **Network Topology** group, click the **Terminal Connections** button to open the **Modify Terminal Connections** pane.

   You'll connect one of the pipes to a terminal. The water main is connected to the **pressure valve device** through terminals. The **Modify Terminal Connections** pane lists **terminals** on the connected devices that are available for assignment.

10. On the map, click the water main line below the **Prssr-Vlv-1** device to select it.

    This line represents **Water Main 2894**.

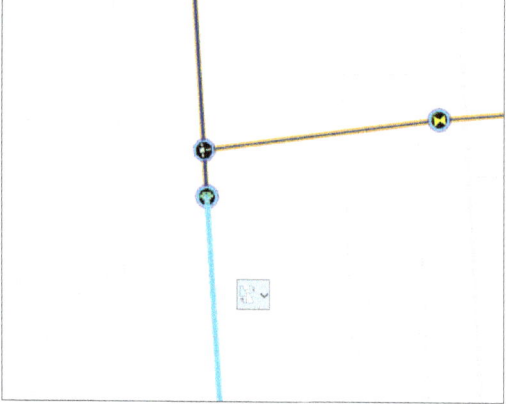

Looking at the water main, the terminal being used to connect the **Pressure Valve** with the transmission main is **High Pressure In**.

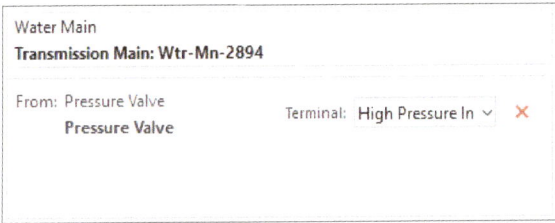

Now that you have a better understanding of terminals and the role they play in connectivity with devices, you can use this knowledge to enable a subnetwork controller for a water pressure zone.

Let's look at subnetwork controllers.

Tutorial 6-2: Set a subnetwork controller

Subnetwork controllers define the origin or termination of a subnetwork (in other words, feeder, pressure zone, or circuit). A controller can be established only on devices whose asset type has been assigned the **Subnetwork Controller** category. For the water distribution pressure tier you're working with, controllers are assigned to the pressure reducing asset type, and the **Pressure Valves** asset group has been assigned to the subnetwork controller category. In your previous tutorial, you worked with **Prssr-Vlv-1**. In this tutorial, you'll set this device as a new controller.

Establish controller name and properties

1. On the **Utility Network** tab, in the **Subnetwork** group, click the **Modify Controller** button to open the **Modify Subnetwork Controller** pane.

   The **Modify Subnetwork Controller** pane allows you to establish name and terminal properties of a controller.

2. In the **Modify Subnetwork Controller** pane, click **Select feature**.

3. On the map, click the **Prssr-Vlv-1** device to select it.

   This new subnetwork controller is set on the low-pressure side of the valve (**Low Pressure Out**).

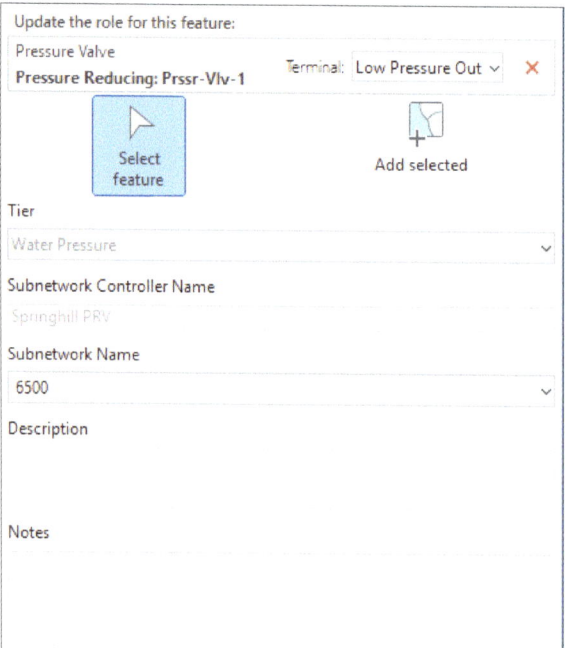

Update subnetwork properties to set the controller

With the subnetwork controller established, the subnetwork is updated with the changes using the **Update Subnetwork** tool. If the subnetwork is newly created, the **Update Subnetwork** tool is used to update the subnetwork name and other propagated values to the connected features.

4. On the **Analysis** tab, in the **Geoprocessing** group, click the **Tools** button to open the **Geoprocessing** pane.

5. In the **Geoprocessing** pane, search for and open the Update Subnetwork tool.

6. In the **Update Subnetwork** tool, apply the following settings:

    - **Input Utility Network**: Water Utility Network
    - **Domain Network**: Water
    - **Tier**: Water Pressure
    - **All subnetworks in tier**: <unchecked>
    - **Subnetwork Name**: 6500

Chapter 6: Terminals and controllers

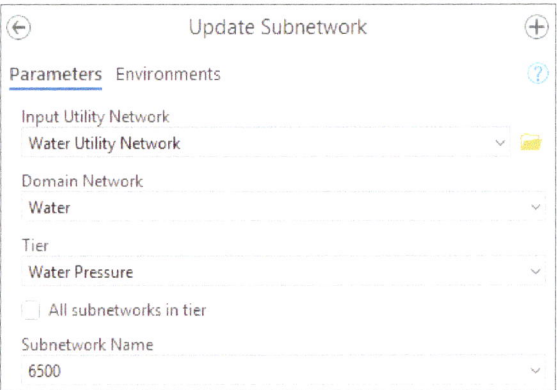

7. Click **Run**.

Update the network controller as needed

If the subnetwork is previously established, the **Find Subnetworks** pane can also be used to update the properties of the feeder after the controller has been modified.

8. On the **Utility Network** tab, in the **Subnetwork** group, click the **Find** button to open the **Find Subnetworks** pane.

9. Locate and right-click **Subnetwork 6500**. Select **Update Subnetwork**.

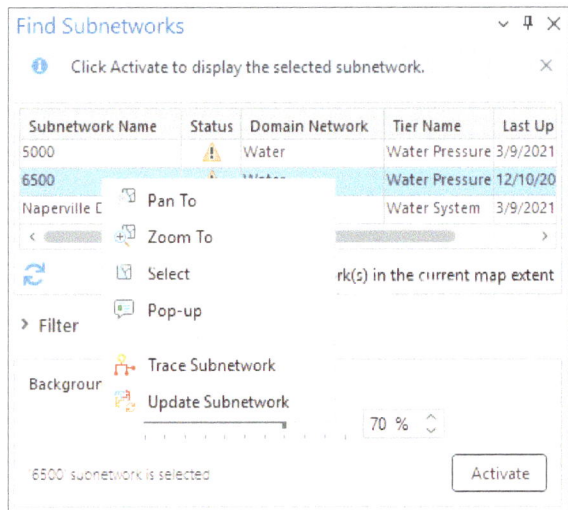

The subnetwork is now updated to include the new subnetwork controller.

On your own

Right-click the **Water Utility Network** layer to browse to the layer properties and identify other network controllers in the system.

Take the next step

Using the **Terminal Paths** tool, limit the available flow paths of a device with terminals.

Summary

In this chapter, you explored how terminals are used on devices to establish appropriate connections with wires and pipes. Terminals are also used to establish subnetwork controllers that act as the source or sink of a subnetwork.

Workflow

1. Identify device terminals.
2. Set a terminal path for a given device.
3. Assign a terminal to a line (pipe) layer.
4. Identify a network controller.
5. Configure a controller and update a subnetwork.

CHAPTER 7
Using network attributes and categories

Objectives

- Describe network attributes and the impacts they have on network topology.
- Describe network categories and how to work with them.

Introduction

In this chapter, you'll learn more about the role of network attributes and categories in utility networks. You'll use network categories and network attributes within different operations in the network.

Network categories allow you to tag specific features within the network, which enhances them with additional functionality. For example, network categories can be assigned to subnetwork controllers or to devices that intersect a line feature midway.

Network attributes use fields from feature and object classes to track characteristics of your network, such as whether a feature is a subnetwork controller. They are stored in the network topology, change as your network does, and are updated when you run **Validate Network Topology** or **Update Subnetwork** on features that have a network attribute assigned.

In later chapters, you'll learn how network categories and network attributes can be used within network tracing and subnetwork management.

Tip: For information about how to set up your utility network project, see the "Getting Started" section. This chapter features the Electric Utility Network Foundation datasets.

Tutorial 7-1: Exploring properties and management of network categories

In this tutorial, you'll learn how to find which network categories are present in your utility network and explore different tools to create, manage, and delete network categories.

Examine network properties

1. From the **Datasets_For_UN_Skills_Book** folder, open **ElectricUtilityNetworkFoundation.aprx**.

2. In the **Electric Network Editor** map, browse to the **Contents** pane.

3. Right-click the **Electric Utility Network** layer and select **Properties**.

4. In the **Layer Properties** window, click the **Network Properties** tab. Scroll down to expand the **Categories** section.

    You can see a list of all categories currently in your utility network.

Name	Created Time
Subnetwork Controller	2023-04-13 13:48:30
Subnetwork Tap	2023-04-13 13:48:30
Attribute Substitution	2023-04-13 13:48:30
Asset Functional	2023-04-13 14:38:12
Asset Location	2023-04-13 14:38:12
Cable Pathway	2023-04-13 14:38:12
Cable Support	2023-04-13 14:38:12
Duct Bank	2023-04-13 14:38:12
Duct Trace	2023-04-13 14:38:12
E:Bank	2023-04-13 14:38:12
E:Cable	2023-04-13 14:38:12

Tip: The network topology must be disabled to add network attributes and network categories.

Create, modify, and remove network categories

It's also important to understand which features within your network make up each network category. A network category is made up of one or more asset group and asset type pairings from a feature class. A single network category can consist of features from different feature classes.

Three tools are available to work with network categories: **Add Network Category**, **Set Network Category**, and **Delete Network Category**. Each tool serves a different purpose, as the following table and images show, but together they give you the ability to create, modify, and remove network categories.

Geoprocessing tools for working with network categories

Tool name	Purpose	Parameters
Add Network Category	Used to add a new network category to the utility network	• Input utility network • Category name
Set Network Category	Used to assign asset groups and asset types to a new network category or modify an existing network category	• Input utility network • Domain network • Input table • Asset group • Asset type • Categories
Delete Network Category	Used to remove a network category from the utility network	• Input utility network • Category name

Note: Network topology must be disabled before running these tools.

Tutorial 7-2: Exploring the composition and impacts of network attributes

In this tutorial, you'll gain a better understanding of how network attributes are assigned within a utility network. You'll also learn how network attributes affect editing.

Examine attributes and assignments

1. In the **Electric Network Editor** map, locate the **Electric Utility Network** layer and open its **Layer Properties** window.

2. On the **Network Properties** tab, scroll down to expand the **Attributes** and **Assignments** sections.

The **Attributes** section lists the names and characteristics of network attributes within the utility network.

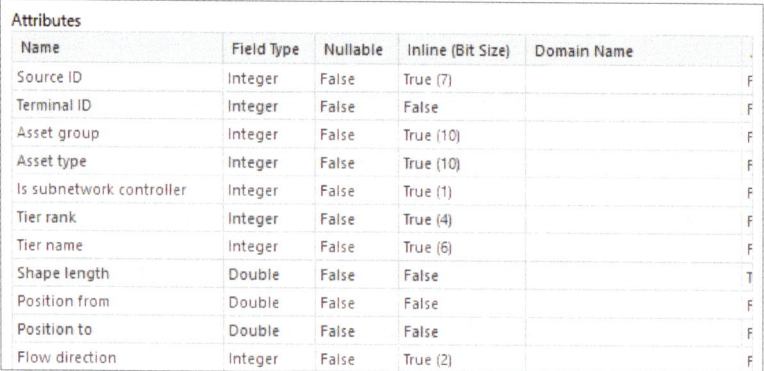

In the **Assignments** section, you can see which fields within feature and object classes are assigned to network attributes. For example, you may want to use a network attribute for the operational status of all devices within your network. To do so, you could create a network attribute called **Status** and assign it to the **Lifecycle Status** field across multiple feature classes.

As you change attributes within these fields, you'll see changes to your network, which you'll learn more about later in the book. Next, you'll explore network attributes within editing to see how the utility network responds when you edit a field that's part of a network attribute compared with one that isn't.

View attributes on a map

3. Close the **Layer Properties** window and return to the map.

4. In the **Contents** pane, expand the **Electric Device** layer.

5. Scroll down within the **Electric Device** layer to the **Medium Voltage Fuse** feature class. Right-click the feature class and select **Attribute Table**.

   This will open the attribute table for the feature class.

6. Select the first record. Scroll to the right until you find the **Lifecycle Status** field header.

7. Confirm that the value for the **Lifecycle Status** of the first record is **In Service**.

8. Above the field headers, in the **Selection** section, click **Zoom To** repeatedly until you can see the feature on the map.

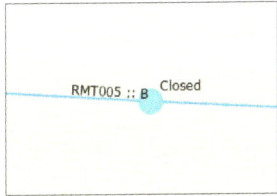

9. Confirm that the feature is selected in the attribute table.

10. On the ribbon, click the **Map** tab. In the **Selection** group, click the **Attributes** button.

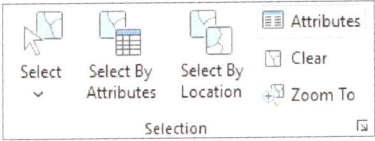

This will open the **Attributes** pane.

As you saw in the **Attributes** and **Assignments** sections of the **layer properties**, not every field within a feature class is part of a network attribute. You'll investigate how editing fields that are not part of network attributes differs from those used in network attributes.

Edit a field

11. In the **Attributes** pane, scroll down through the **Attributes** list to locate the **Manufacturer** field.

12. Change the value of the **Manufacturer** field to ERM Co and click **Apply**.

    The attribute is now updated.

    > **Tip:** If you turn on the Auto Apply toggle within the Attribute pane, your edits will automatically be made to the feature without the need to click Apply.

    From reviewing the **Assignments** group in the **Electric Utility Network** layer properties earlier in the tutorial, you know that the **lifecyclestatus** field within the **electric device** is assigned to a network attribute called **Lifecycle Status**.

Lifecycle Status	StructureLine	lifecyclestatus
Lifecycle Status	ElectricAssembly	lifecyclestatus
Lifecycle Status	ElectricDevice	lifecyclestatus
Lifecycle Status	ElectricJunction	lifecyclestatus

    Next, you'll examine the impact that editing a network attribute has on the network topology.

13. In the **Attributes** pane, change the value of the **Lifecycle Status** field to **Proposed**. Click **Apply** and zoom in on the feature.

    You should see a box around the feature that wasn't present before. This box was created because you edited a network attribute on the feature. Network attributes are stored in the network topology itself, so your edit changed the state of topology.

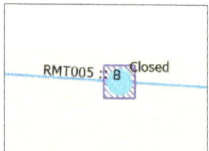

On your own

Investigate the preconfigured network categories and network attributes in the data to gain a better understanding of what parts of the data model use them.

Take the next step

Review your organization's data to evaluate where you could use network categories to group features and what fields within your data you may want to use for network attributes.

Summary

In this chapter, you learned more about network attributes and network categories.

Workflow

1. Examine the geoprocessing tools for working with network categories: Add Network Category, Modify Network Category, and Delete Network Category.
2. Locate a feature and edit a field that isn't part of a network attribute.
3. Edit a field that participates in a network attribute.
4. Examine the changes to the network that result from modifying a field that's part of a network attribute.

CHAPTER 8
Applying network rules and validating data

Objective

- Examine utility network rules and the role they play in data validation and correctness.

Introduction

In this chapter, you'll learn about utility network connectivity, containment, and attachment rules and explore the role they play in data validation as you make edits.

Network rules play a critical role in the editing process by providing a framework of logical rules that determine what types of features can be connected, how features can be associated together or contained within one another, and how features can be attached. As edits to features and associations are made, they are continually checked against the network rules to ensure that the edits being made align with the rules. When edits conform to the rules of the network, a dirty area is created. When edits don't follow the rules of the network, an error feature is generated indicating that a rule has been violated by the edit.

In the tutorials for this skill, you'll examine the utility network rules and then see how they work within the editing process to help you create correct data.

Chapter 8: Applying network rules and validating data

Tip: For information about how to set up your utility network project, see the "Getting Started" section.

Tutorial 8-1: Examining different types of network rules

In this tutorial, you'll examine the utility network properties to gain a better understanding of the rule types for connectivity, attachment, and containment and dive into what a rule is made up of.

Locating rules in the utility network

You'll open the utility network properties and locate the **Rules** group on the **Network Properties** tab.

1. From the **Datasets_For_UN_Skills_Book** folder, open **ElectricUtilityNetworkFoundation.aprx**.

2. In the **Electric Network Editor** map, browse to the **Contents** pane.

3. Right-click the **Electric Utility Network** layer and select **Properties**.

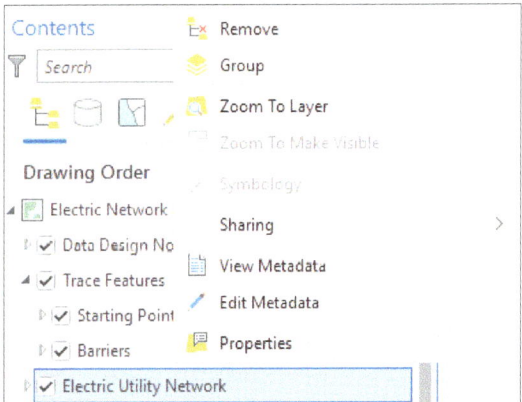

4. In the **Layer Properties** window, click the **Network Properties** tab.

5. Scroll down to expand the **Rules** section.

```
∨ Rules
    > Junction-Junction Connectivity
    > Junction-Edge Connectivity
    > Edge-Junction-Edge Connectivity
    > Containment
    > Structural Attachment
```

The **Rules** section of the utility network properties contains all the information about the configuration of connectivity, containment, and attachment rules in your utility network.

Network rules, regardless of whether they are connectivity, containment, or attachment, define the only types of valid connections, associations, or attachments. If a rule doesn't exist to tie features together, the system will create an error feature in the dirty areas table, with additional information regarding why the edit is invalid and not make the change to the data.

Connectivity rules define what types of spatial features can be connected to one another. Junction-junction connectivity group contains rules for how two-point features, or junctions, as they're known in the utility network, can be connected. The junction-edge connectivity group holds rules for how a point can be connected to a line, or edge, as they are referred to in the utility network. Finally, the edge-junction-edge connectivity group includes rules for how more complex connectivity between two line features and a point feature can be connected.

Containment rules define what features can be nested within one another as containers and content. Containers and content exist within a hierarchy, with the container feature being the parent feature and the content being the child. For example, a pump station in a water system could be used as a container, and each piece of equipment within the station could be associated individually as content.

Structural attachment rules govern how features can be associated with one another within the structural domain network. For example, an electric user could use the structural attachment rules to define what types of features could be associated with a pole feature.

Edge (line) to junction (point) rule type

Edge to junction to edge rule type

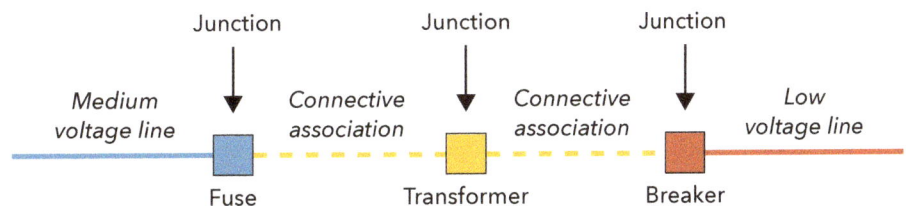

Junction (point) to junction (point) rule type

Containment rule type

Containment view / large scale

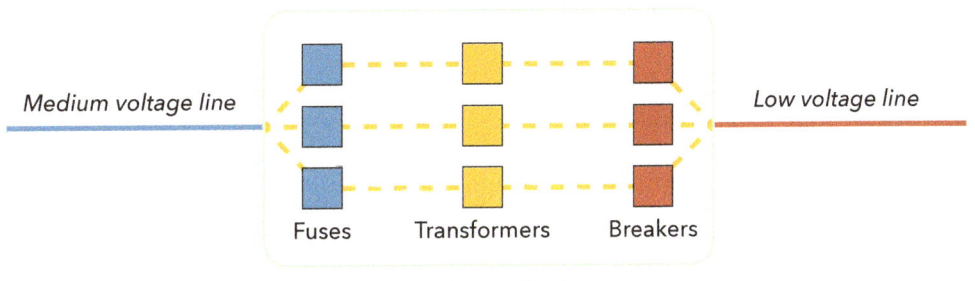

Simple view / small scale

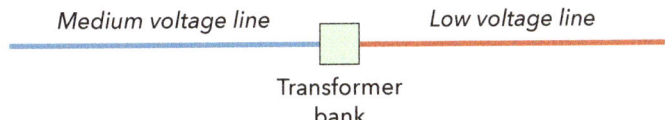

Structural attachment rule type

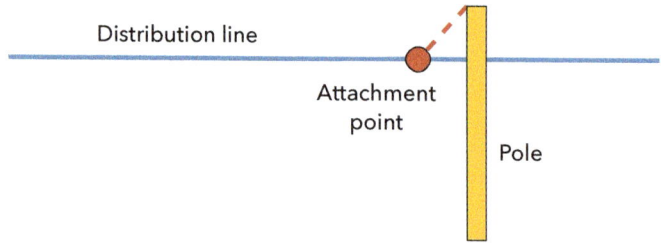

Together, the rules provide a logical framework to ensure that correct data is being created as you edit.

6. Expand the **Junction-Edge Connectivity** group.

Tip: Network rules contain three common components: a feature class, an asset group, and an asset type. The Esri foundation models for each utility domain contain many rules within them that have been designed to match how devices and infrastructure are connected in the real world.

7. Close the **Layer Properties** window.

In this tutorial, you learned more about connectivity, containment, and attachment rules. In the next tutorial, you'll see the impact of network rules in data validation as you edit features.

Chapter 8: Applying network rules and validating data

Tutorial 8-2: Working with network rules in editing

In this tutorial, you'll create features and make edits to understand how network rules and validation help you create correct data.

Create features and make edits

You'll create an underground conductor and service point to gain insights into how network rules validate your edits.

1. On the ribbon, click the **Map** tab. In the **Inquiry** group, click the **Locate** tool to open the **Locate** pane.

2. In the **Locate** pane, click the **Layer Search** tab and search for a device with an **asset ID** of MV-XFR-BK-1244.

3. In the results, right-click the feature and click **Zoom To**.

4. On the ribbon, click the **Edit** tab. In the **Features** group, click **Create** to open the **Create Features** pane.

5. In the **Create Features** pane, search for Low Voltage Underground Conductor.

6. Within the **Electric Line : Low Voltage Underground Conductor** template, click **AC Underground LV**.

7. On the map, with the **Line** tool selected, hover over the square **Medium Voltage Transformer** feature.

    You should see the cursor snap to the feature instantly.

    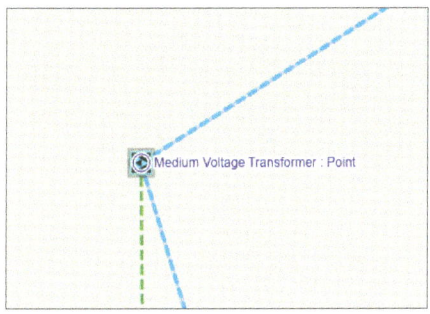

When you're editing, the cursor snaps to the **Medium Voltage Transformer** feature instantly because the features being connected adhere to junction-edge connectivity rules. If no rule existed, the system wouldn't snap features as shown earlier.

8. Sketch the conductor line feature around the edges of the building and connect it to the wall.

9. Save your edits and run the **Validate Network Topology** tool. Refer to the following image.

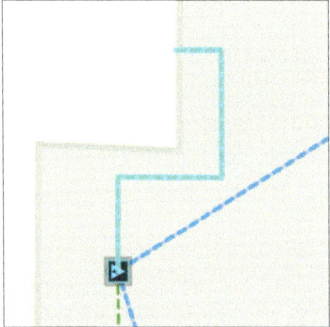

10. In the **Create Features** pane, search for Service Point.

11. Scroll down to the **Electric Device : Low Voltage Service** feature template and select **Three Phase Residential LV**.

12. Snap the service point to the end of the line you created in step 8.

13. Save your edits and run the **Validate Network Topology** tool.

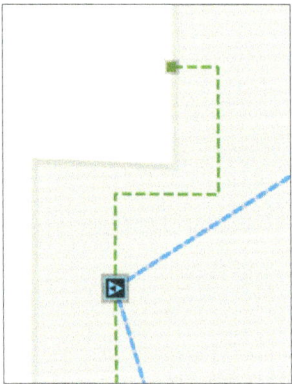

On your own

Study network rules within the foundation data to better understand how they're used to represent connections between infrastructure in the real world. Use this knowledge to begin planning out what rules you may want to add that are specific to your organization.

Take the next step

Read about the **Add Rule** geoprocessing tool in the Esri Help documentation to gain a deeper understanding of how you can create your own custom rules. Go to pro.arcgis.com, click the Help menu, click Tool Reference > Geoprocessing Tools > Utility Network Toolbox, and scroll down to the Administration toolset link.

Export the rules from the foundation data using the **Export Rules** geoprocessing tool. You can use the comma-separated value file in external applications or share it with coworkers.

Summary

In this chapter, you learned about connectivity, containment, and attachment network rules as well as how network rules affect the editing experience. You also created and then resolved an error feature in your network.

Workflow

1. Review the Rules section of the utility network properties to understand the rule configuration of the network.
2. Use the Locate tool to navigate to the device.
3. Create a service line and service point.
4. Use the Validate Network Topology tool.
5. Save your edits.

CHAPTER 9
Using the error inspector to resolve error features

Objectives

- Describe error features in the utility network.
- Use the error inspector to locate error features.
- Use basic editing skills and knowledge of network rules to correct error features.

Introduction

Data quality control for managing a utility network is critical for performance and usability. Network rules enable greater enforcement during edits, but there are still occasions when network data is in an error state. To help identify, locate, and correct errors, the error inspector was modified to support error messages in the **Utility Network** layer. Although many volumes have been written about best practices of correcting errors (including modifying rules or running batch processes), this chapter will use simple edits of network attributes to demonstrate a feature in an error state and then correct and validate to show how the errors are cleaned up.

> **Tip:** For information about how to set up your utility network project, see the "Getting Started" section. This chapter features the Gas and Pipeline Utility Network Foundation dataset.

Tutorial 9-1: Exploring the error inspector

Open the error inspector

1. From the **Datasets_For_UN_Skills_Book** folder, open **Gas and Pipeline Utility Network.aprx**.

2. In the **Gas and Pipeline Network** editor map, browse to the **Utility Network** tab. In the **Network Topology** section, click the **Error Inspector** tool.

   The main section of the pane displays the error details on the left and the feature details on the right. Because the **Gas and Pipeline Foundation** dataset has been scrubbed for errors, we have no issues to display. You'll create an error feature to explore the error inspector.

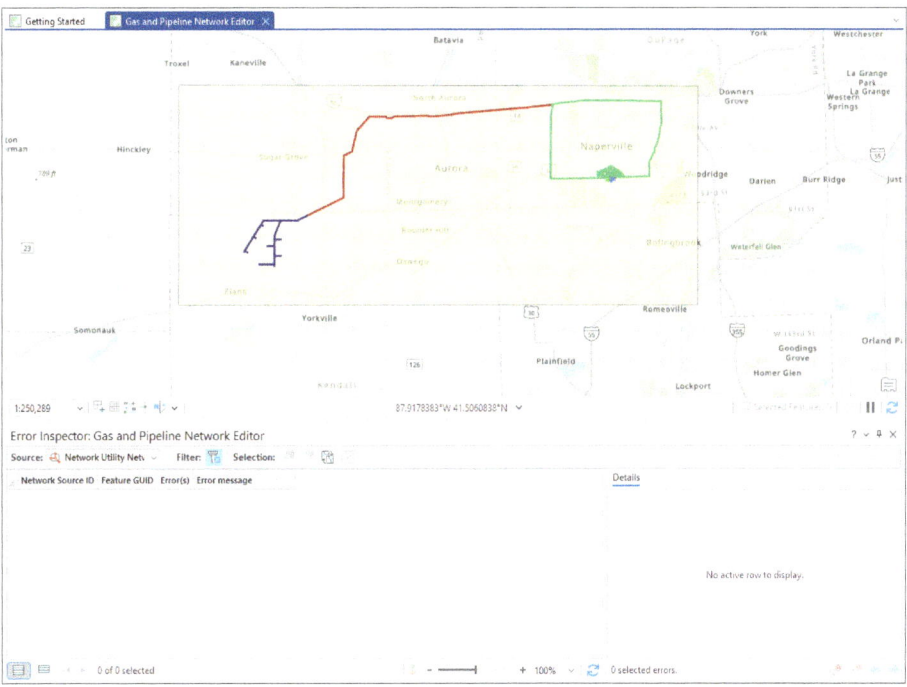

Locate a feature to test with the error inspector

3. On the ribbon, click the **Map** tab. In the **Inquiry** group, click the **Locate** button to open the **Locate** pane.

   You'll test a service pipe with an **asset ID** of **SVC-Pp-346**.

4. In the **Locate** pane, click the **Layer Search** tab and search for SVC-Pp-346.

5. In the results, right-click the service pipe and click **Zoom To** until you can see the individual feature on the map.

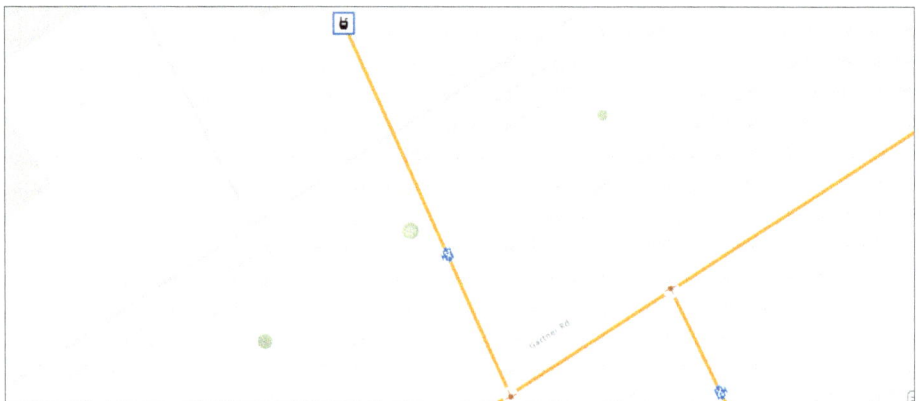

Change the asset type of the selected feature to create an error feature

In chapter 8 ("Applying Network Rules and Validating Data"), you learned the importance of network rules in the network index. This section shows what happens if features on the map violate the network rules.

6. On the **Map** tab, in the **Selection** group, click the **Attributes** tool to open the **Attributes** pane.

7. In the **Attributes** pane, click **Select one or more features**.

8. On the map, click the **SVC-Pp-346** service pipe to select it.

9. In the **Attributes** section of the pane, find the **Asset type** field.

Chapter 9: Using the error inspector to resolve error features

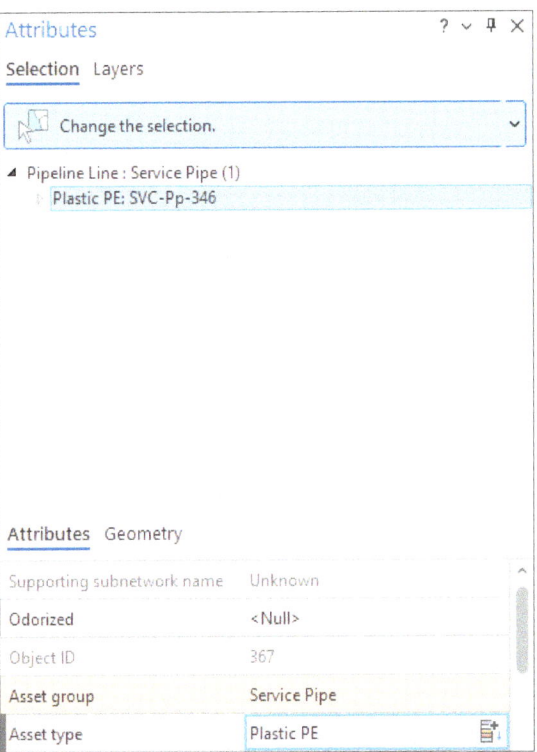

Create the dirty area for the edited feature

10. Change the **Asset type** field from **Plastic – CE** to **Unknown**.

11. Click **Apply**.

    Upon completion of the edit, a dirty area is generated around the edited feature.

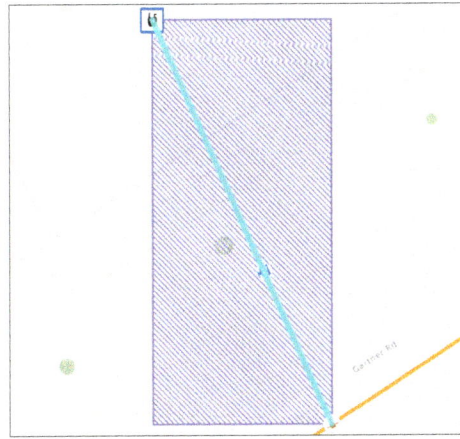

Create an error feature

Running a validation on the dirty area creates an error feature that represents the network features that violate the network rules in the error inspector.

12. Click the **Utility Network** tab. In the **Network Topology** group, click the **Validate** button to run the **Validate Network Topology** tool.

13. Review the changes in the **Error Inspector** pane.

    The left side of the **Error Inspector** pane provides details about the errors in the display, whereas the right side shows attribute details. The feature is in an error state because the asset type value of **Unknown** doesn't have network rules that allow for midspan connectivity, and they don't allow for the connections on both ends of the line.

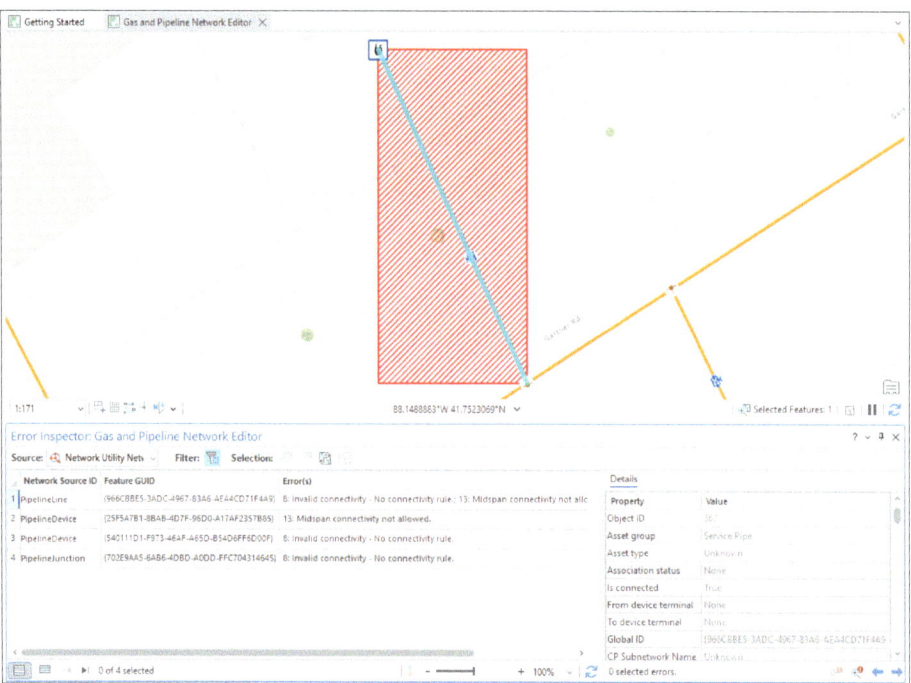

Repair an error feature

Using the error inspector, you can scroll through the list of features in error and examine why they're in error. Rule-driven issues can be corrected by changing attributes to correct values (such as the preceding example), creating terminal connections (as discussed in chapter 6) or offsetting features. These are just a few examples of how errors can be located and corrected using the error inspector.

14. In the **Attributes** pane, restore the service pipe error feature by changing the **Asset type** value from **Unknown** to **Plastic CE**.

On your own

Use your knowledge of network rules from chapter 8 to see whether you can build more service laterals in the gas system.

Take the next step

Use the error inspector to correct any errors created during your tutorials when learning network rules.

Summary

The error inspector enables quick identification and diagnostics of features in an error state. As you work through tutorials in the other **Foundation** datasets, feel free to use the error inspector to help you clean up errors in your system.

Workflow

1. Locate a service lateral.
2. Change the asset type to create an error feature.
3. Validate to create an error.
4. Use the error inspector to see error messages.

CHAPTER 10
Creating trace locations

Objectives

- Learn how to create starting points.
- Learn how to create barriers.

Introduction

In the next few chapters, you'll explore different trace types, learn more about a variety of trace parameters, and discover how you can visualize and use trace results. In this chapter, you'll learn about the role of starting points and barriers in the utility network and explore different methods to create them.

Starting points and barriers are considered trace locations. Starting points are the origin points from which a trace emanates. Conversely, barriers restrict the path of a trace in the network. Starting locations and barriers are also referred to as trace locations. Starting points and barriers can be placed on lines and point features, which gives you the flexibility to start and stop your traces where you want.

Starting points and barriers are derived from a selection set of features or manually placed within the UI and can be saved for future use in geodatabases.

Certain trace types, such as connected, upstream, downstream, and shortest path, require starting points to be executed.

Tip: For information about how to set up your utility network project, see the "Getting Started" section. This chapter features the Electric Utility Network Foundation dataset.

Tutorial 10-1: Creating starting points

In this tutorial, you'll learn different methods to create starting points and barriers in the utility network. You'll also gain a conceptual understanding of starting points and barriers to prepare for later skills, where you'll concentrate on network tracing.

Use a feature selection to create a starting point

In this section, you'll learn how to create a starting point using the **Trace** pane and a selection of features.

1. From the **Datasets_For_UN_Skills_Book** folder, open **ElectricUtilityNetworkFoundation.aprx**.

2. In the **Electric Network Editor** map, click the **Map** tab. In the **Inquiry** group, click the **Locate** button to open the **Locate** pane.

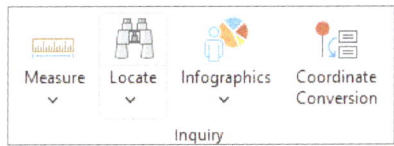

3. In the **Locate** pane, click the **Layer Search** tab and search for a breaker with an **asset ID** of MV-CB-18.

   Next, you'll place a starting point.

4. Right-click the first result, **Three Phase Circuit Breaker MV–Controller: MV-CB-18**, and click **Add To Selection**.

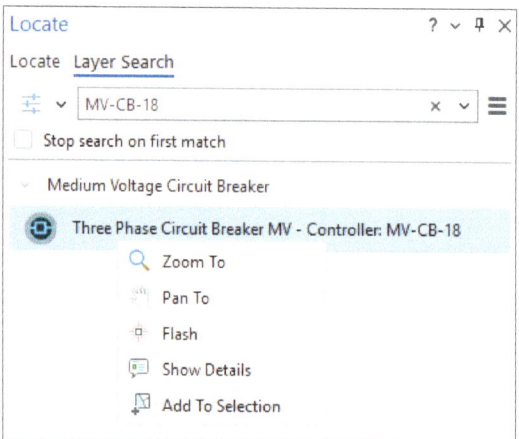

5. Right-click **MV-CB-18** again and click **Zoom To**. Adjust the zoom so that you can see the features individually.

   Tip: When you click Zoom To, the map will display features at the extent to which the map scale is set. If you want to see a feature up close, set the map scale in the bottom-left corner to 1:100 or 1:500. To adjust the map scale, you can use one of the predefined map scales in the list or set your own by typing an integer in the Scale box.

The **MV-CB-18** device is now selected and zoomed to in the map.

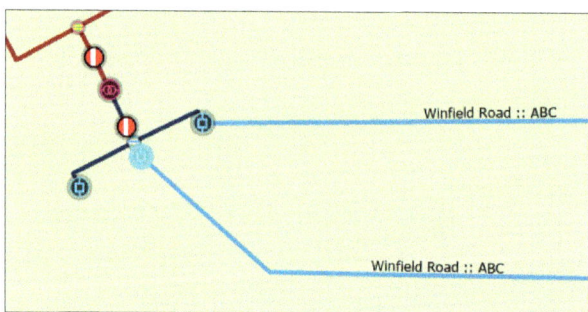

   Tip: The Locate tool can be used to search for features on which you want to place starting points or barriers. Instead of searching attribute tables, you can query the feature you're looking for.

Next, you'll place a starting point on the selected feature.

6. With the device still selected, click the **Utility Network** tab. In the **Tools** group, click the **Trace** button to open the **Trace** pane.

7. With the **Trace** pane open and the **Start** tab selected, click the **Add selected** button.

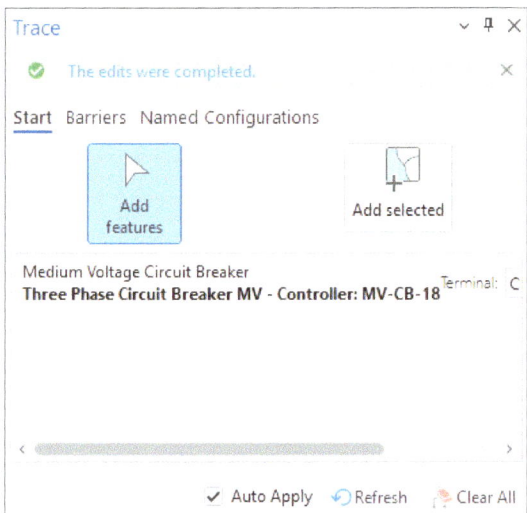

The selected feature will now show as a starting point in the **Trace** pane.

8. Confirm that the feature on the map also has a green circle coincident with the device, which indicates the presence of a starting point.

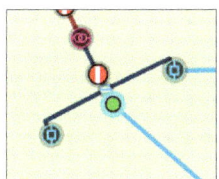

Starting points can also be added from the **Trace** pane by clicking the **Add features** button and then clicking a feature on the map. A green starting point will appear.

> **Tip:** If you need to remove trace locations from the map, you can use the Clear All button on the Trace pane to remove starting points or barriers.

9. On the **Map** tab, in the **Selection** group, click **Clear** to clear your selection.

You have now successfully created a starting point using a selection set and the **Add features** button on the **Trace** pane. In the next tutorial, you'll add barriers.

Tutorial 10-2: Creating barriers

Barriers are created in a similar fashion. You can click a feature to add a starting point to it or use a selected set of features as starting points.

Use a feature selection to create a barrier

1. In the **Locate** pane, use the **Layer Search** tab to search for a terminator with an **asset ID** of MV-LE-901.

    You'll place a barrier on this device.

2. Right-click the first result, **Underground Terminator MV: MV-LE-901**, and click **Add To Selection**.

3. Adjust the zoom so that you can see all the features individually.

4. On the map, click the **MV-LE-901** device to open its **Pop-up** window.

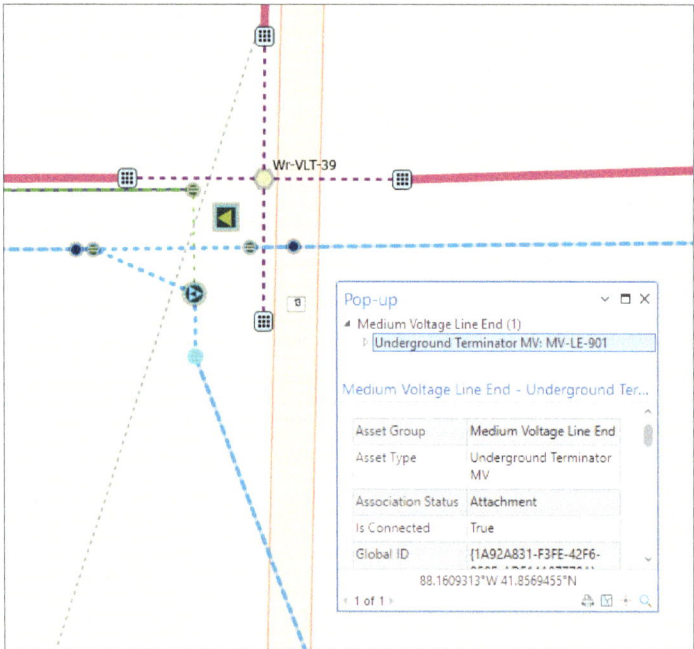

5. In the **Trace** pane, click the **Barriers** tab.

6. Click the **Add selected** button to add a barrier to the map.

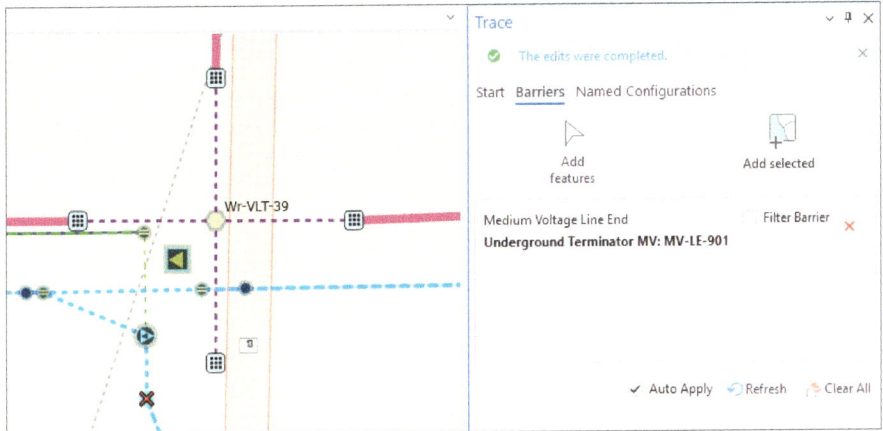

7. Barriers can also be added using **Add features** in the **Trace** pane. The *X* atop the device, shown in the image, represents a barrier in the network.

   Tip: Clicking Clear All will remove starting points and barriers.

8. In the **Trace** pane, click **Add features** and then click any point or line feature.

   Your new barrier selections will be listed in the **Trace** pane.

On your own

Investigate your organization's data to determine what features on which you want to place starting points or barriers.

Take the next step

Learn more about how you can save starting points and barriers for reuse in the geodatabase.

Summary

In this chapter, you learned more about the role of starting points and barriers in the utility network and gained an understanding of how to create them.

Workflow

1. Search for a feature using the Locate tool.
2. Zoom to the feature and confirm that it's selected.
3. Use the Trace pane to add starting points and barriers using one of the following methods:
   - Using a selection set of features.
   - Using the Add Feature tool and clicking a feature.

CHAPTER 11
Using a basic connected trace

Objectives

- Use starting points and barriers.
- Generate a selection set of features from a trace.
- Describe the results of a trace.
- Introduce the elements of the trace framework.

Introduction

For utilities, most assets are linear, such as pipes or wire and cable. Many daily activities involve locating features or points down the map line. In some cases, those linear queries are stopped based on the state of a feature or a numeric value attributed to a feature on the line.

The connected trace is a specialized query that uses the network index to return a set of features that are linked by geometric coincidence or association (junction-to-junction, containment, or structural). A starting point is created, and the resulting trace extends in all directions along connected features; the trace includes all features until it reaches the end of the network or encounters a barrier (feature or dynamic).

Of the trace types, the connected trace is the most basic. It's not restricted in terms of direction and can be used to address a range of spatial problems.

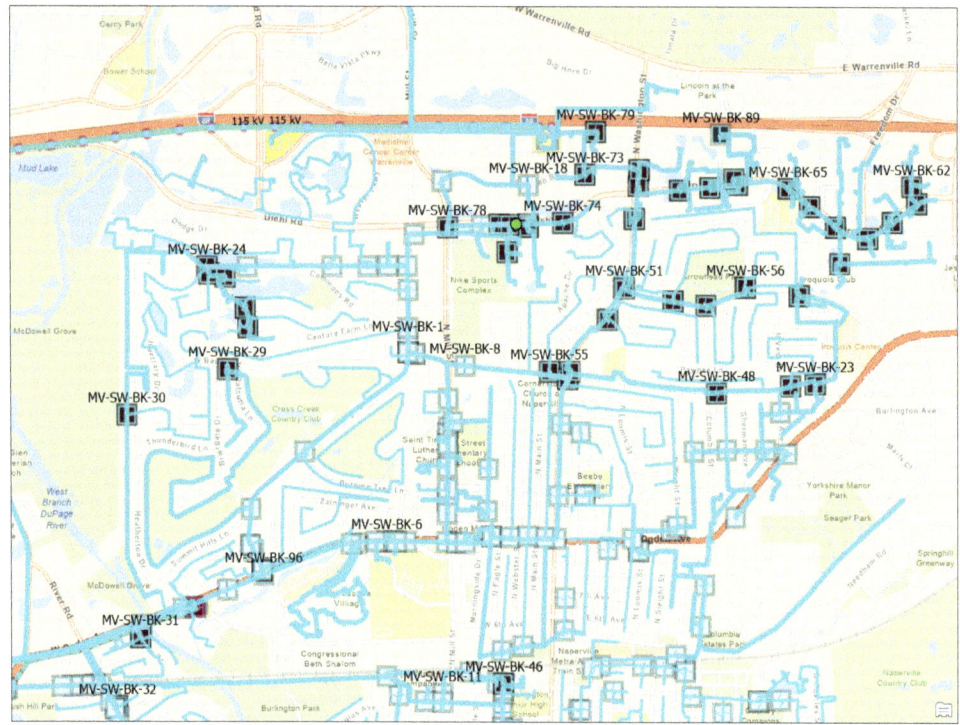

Tip: For information about how to set up your utility network project, see the "Getting Started" section. This chapter features the Electric Utility Network Foundation datasets.

Tutorial 11-1: Configuring and running a connected trace

In this tutorial, you will learn how to configure and run a connected trace.

Work with the trace framework

1. From the **Datasets_For_UN_Skills_Book** folder, open **ElectricUtilityNetworkFoundation.aprx**.

2. In the **Electric Network Editor** map, on the bottom-left corner of the map, set the **Scale** to 1:1000.

Chapter 11: Using a basic connected trace   83

3. Open the **Locate** pane. Click the **Layer Search** tab and search for a streetlight with an **asset ID** of LV-LGHT-1019.

   This feature is a streetlight where you will place your starting point.

4. Right-click the returned value, **Streetlight LV: LV-LGHT-1019**, and click **Add To Selection**.

5. Adjust the zoom so that you can see the **LV-LGHT-1019** streetlight highlighted on the map.

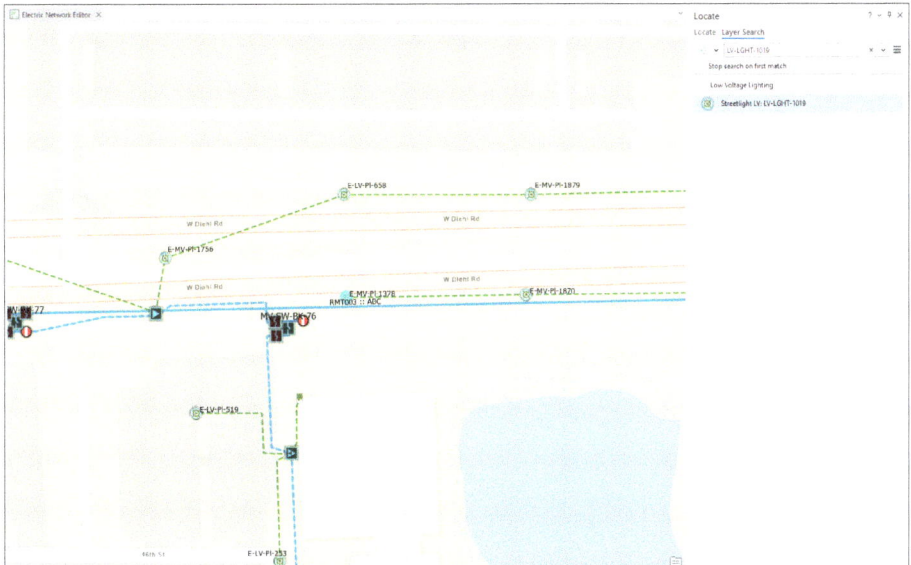

6. On the **Utility Network** tab, open the **Trace** pane. Click the **Start** tab.

7. Use the **Add Features** tool to click the blue line between streetlight pole 1378 (**E-MV-Pl-1378**) and 1870 (**E-MV-Pl-1870**).

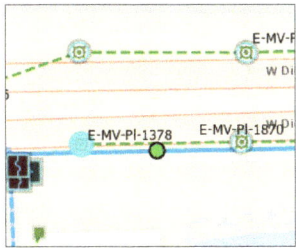

8. When you have placed the starting point, click the **Utility Network** tab. In the **Tools** group, click the **Connected** tool.

The **Geoprocessing** pane opens, showing the **Trace** tool, which has been configured for a connected trace.

For this connected trace, you'll keep the default values for all parameters. In later chapters, you'll learn how to customize traces to modify parts of the network.

9. In the **Geoprocessing** pane, click **Run**.

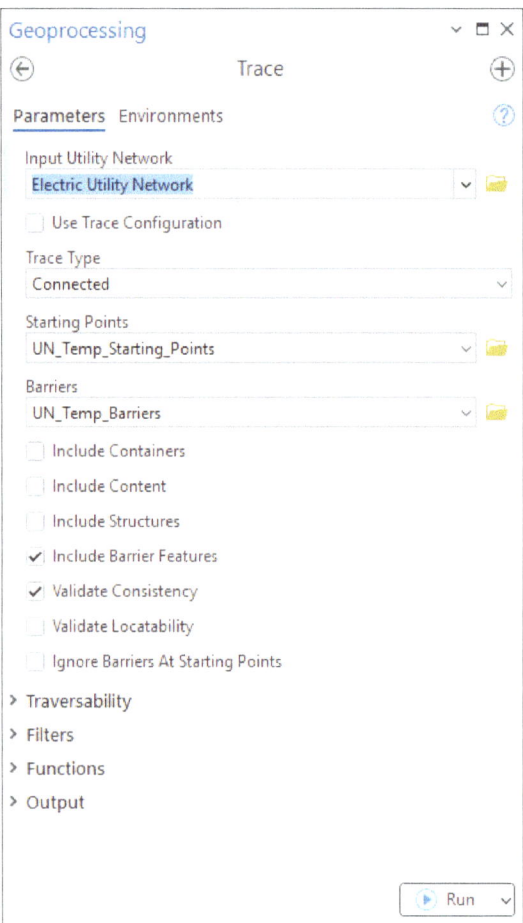

A selection is returned of all features connected to the line on which the starting point resides.

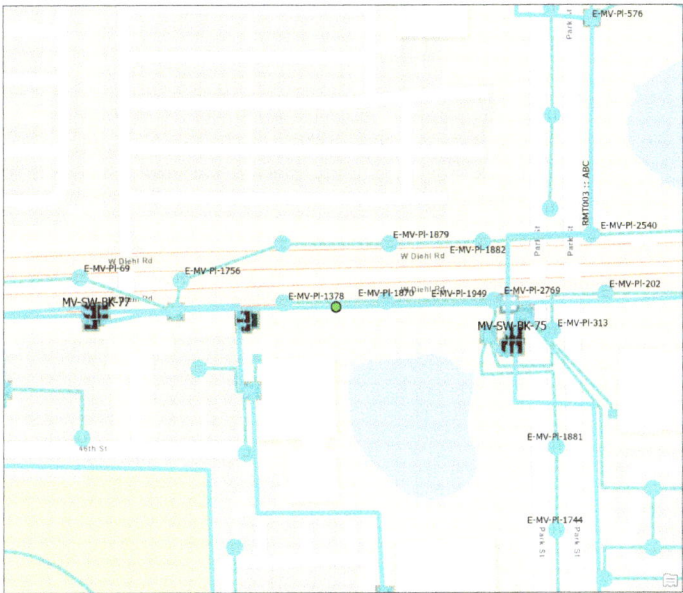

On your own

Identify features within your network where you may want to place starting points or barriers for use in connected traces.

Take the next step

Test tracing in the utility network with and without dirty areas or network errors to see how they affect tracing.

Summary

In this chapter, you learned how connected traces enable performant spatial queries along network features. This skill introduces the trace framework and lays the foundation for more complex trace configurations over the following chapters.

Workflow

1. Open the project.
2. Place a starting point on the line.
3. Run the connected trace.
4. Review the results.

CHAPTER 12
Using directional traces

Objective

- Use a directional trace such as an upstream or downstream trace.

Introduction

In this chapter, you'll learn more about how to run directional traces and how to interpret the results of the trace.

Utilities of all types need to understand how resources flow through their infrastructure for operational, safety, and maintenance purposes. These resources, whether they are water, gas, or electricity, all originate somewhere and are deposited somewhere else within the network.

The utility network has been designed to anticipate these needs by tracking the direction that resources can flow and allowing you to use that capability as you trace.

These traces use the directionality stored by the utility network within the network topology to return a selection set of features that are linked by geometric coincidence or association (connectivity, containment, or structural attachment), which are upstream or downstream of a starting point in the utility network.

Tip: For information about how to set up your utility network project, see the "Getting Started" section. This chapter features the Electric Utility Network Foundation datasets.

Tutorial 12-1: Working with a downstream trace

In this tutorial, you'll be introduced to the configuration parameters of a downstream trace and learn how to interpret the trace results.

Open a feature

1. From the **Datasets_For_UN_Skills_Book** folder, open **ElectricUtilityNetworkFoundation.aprx**.

2. In the **Electric Network Editor** map, on the bottom-right corner of the map, set the scale to 1:1,000.

3. Open the **Locate** pane. Search for a pipe feature with an asset ID of MV-COND-402. Add it to your selection and zoom to the line on the map.

   Tip: For further information on how to use the Locate tool or adjust the map scale, see chapter 10 ("Creating Trace Locations").

4. Open the **Trace** pane. Use the **Trace** tool to create a starting point on the midpoint of **MV-COND-402**, as shown in this image.

   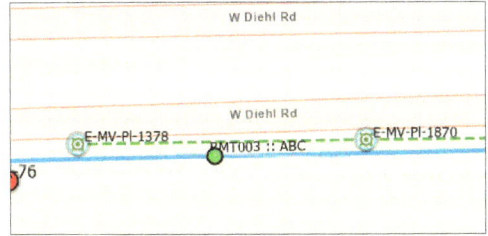

5. Click the **Utility Network** tab. In the **Tools** group, click the **Downstream** tool.

   The **Geoprocessing** pane with the **Downstream** tool will open preconfigured for a downstream trace.

   Tip: You may need to expand the Tools list to find the Downstream tool.

6. In the **Geoprocessing** pane, keep the default values for all parameters except for the following modifications:

   - **Domain Network**: Electric
   - **Tier**: Electric Distribution
   - **Target Tier**: Electric Secondary

7. Click **Run** to run the **Trace** tool.

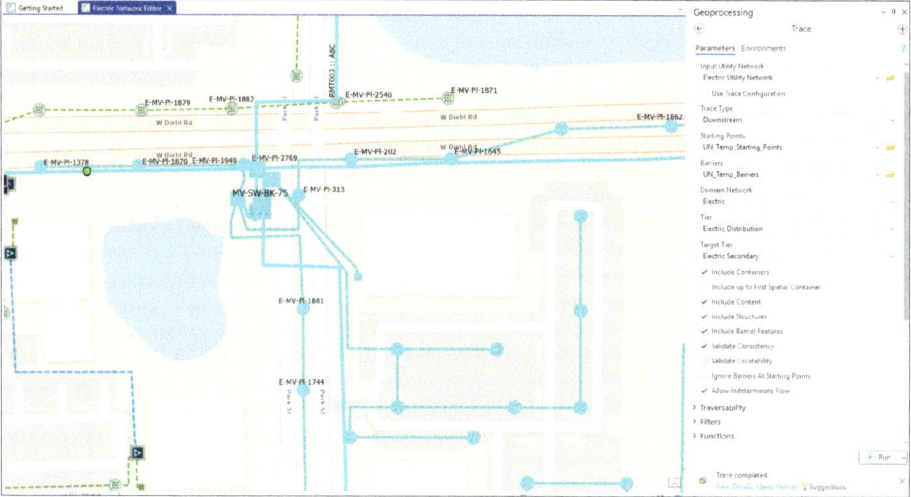

Now you have performed a downstream trace. You have used the network topology to find all features downstream of where your trace began, the starting point on the map.

> **Tip:** When you run a trace, the starting points and barriers default to using the starting points and barriers feature classes in the geodatabase. The parameters can be modified to use object classes in the network as well as asset groups in feature classes in the domain or structure network.

Tutorial 12-2: Working with an upstream trace

Upstream traces enable performant spatial queries along network features. This tutorial introduces the trace framework and lays the foundation for more complex trace configurations over the following chapters. This chapter builds off the downstream trace you ran in tutorial 12-1 by using the same starting points.

Configure the Trace tool

1. On the **Map** tab, in the **Selection** group, click **Clear** to clear your selection from the downstream trace.

2. Focus your map extent on the starting point placed on **MV-COND-402** from tutorial 12-1.

3. Confirm that the starting point is present in the map with a green circle.

4. On the **Utility Network** tab, in the **Tools** group, click the **Upstream** tool.

   The **Geoprocessing** pane with the **Trace** tool opens and is configured for an upstream trace.

5. In the **Geoprocessing** pane, keep the default values for all parameters except for the following modifications:

   - **Domain Network**: Electric
   - **Tier**: Electric Distribution
   - **Target Tier**: Electric Transmission

6. Click **Run**.

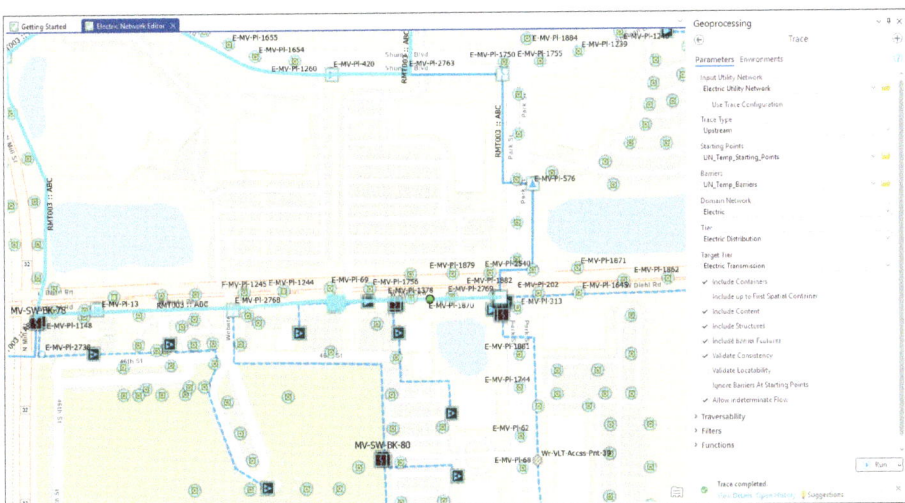

Now you have performed an upstream trace. Like the downstream trace, the upstream trace uses the directionality within the utility network topology to help locate features upstream of your starting point.

> **On your own**
>
> Place a starting point or barrier in different locations within the network and perform an upstream or downstream trace to gain a better understanding of directionality within the utility network.

Take the next step

Explore **Trace** tool geoprocessing parameters such as **Include Structures**, **Include Content**, and **Include Containers**. Enabling or disabling these parameters can change the trace results considerably. Running traces with and without these options gives you a better understanding of what features are part of the structure network versus the domain network, as well as what features are in containment.

Use specific asset groups within domain or structure feature classes as starting points or barriers and see how this affects your trace.

Summary

In this chapter, you learned more about directional upstream and downstream traces and how to run them using the **Trace** geoprocessing tool.

Workflow

1. Place a starting point.
2. Assign domain, tier, and target tier to meet trace requirements.
3. Run the Trace tool.

CHAPTER 13
Applying function barriers

Objectives

- Learn how to create a function barrier.
- Learn how a function barrier affects a trace.

Introduction

In the next few chapters, you'll learn about function barriers and filter function barriers and the impact they have on trace results.

Function barriers allow users to control how far from the starting point the trace traverses using numeric or logical conditions. For example, users could configure an upstream trace to traverse 100 feet using the shape length of the lines within the utility network.

Function barriers in the utility network all conditionally restrict the path of the trace, but each type works a bit differently.

Time to trace.

> **Tip:** For information about how to set up your utility network project, see the "Getting Started" section. This chapter features the Water Distribution Utility Network Foundation datasets.

Tutorial 13-1: Understanding and creating function barriers

Function barriers can be helpful in locating features within a specific portion of a subnetwork. If an electrical fault occurs or a water main breaks, a function barrier could be used to generate a selection of customers affected by the incident.

In this tutorial, you'll configure a downstream trace with a function barrier. This type of trace is helpful for limiting the trace results to a specific part of the network while also using the directionality within the utility network.

Locate the area of interest and create the function barrier

1. From the **Datasets_For_UN_Skills_Book** folder, open **Water Distribution Utility Network Foundation.aprx**.

2. In the **Water Distribution Editor** map, open the **Locate** pane.

3. In the **Locate** pane, click the **Layer Search** tab and search for a feature with an **asset ID** of Wtr-Mn-4154.

4. Right-click the result, click **Add To Selection**, and then click **Zoom To** in order to navigate to the feature.

   You have now located the line feature where the issue is occurring.

5. Place a starting point on **Wtr-Mn-4154**.

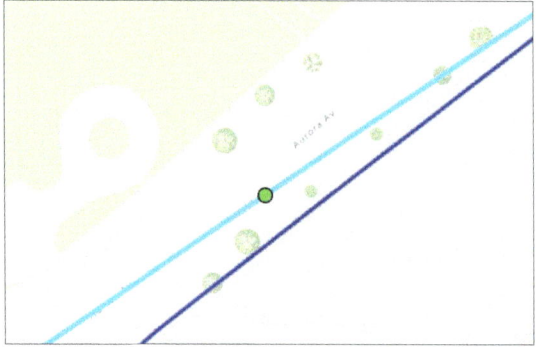

The issue is affecting infrastructure 8,000 feet downstream of the line feature where you placed the starting point. A downstream trace using a custom function barrier can help locate all affected devices.

Chapter 13: Applying function barriers

6. In the **Geoprocessing** pane, use the **Trace** tool to configure a downstream trace. Keep the default values for all parameters except for the following modifications:

   - **Domain Network**: Water
   - **Tier**: Water System
   - **Target Tier**: Water Isolation

7. Expand the **Traversability** group and review the **Function Barriers** section.

   You'll use a function to create a barrier after the trace has reached 8,000 feet downstream of the starting point. Because you're using the **Add** parameter, the trace will sum the shape length of all lines until it reaches 8,000. When the trace reaches this point, it will terminate.

8. Apply the following modifications to the **Function Barriers** parameters. Keep the default values for all other values:

   - **Function**: Add
   - **Attribute**: Shape length
   - **Operator**: Is greater than or equal to
   - **Value**: 8000

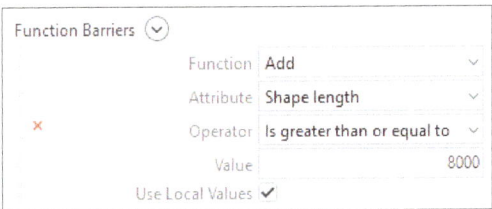

Tip: The coordinate system of the Water Distribution Utility Network Foundation data is NAD 1983 StatePlane Illinois East FIPS 1201 (US Feet), which uses feet as the linear unit. As a result, the distance entry within the function barrier parameters will be in feet. If the linear unit of the coordinate system were meters, the function barrier would calculate the distance in meters.

9. Click **Run**.

   After the **Trace** tool completes, the map shows a selection set of all features that are within 8,000 feet of the starting point.

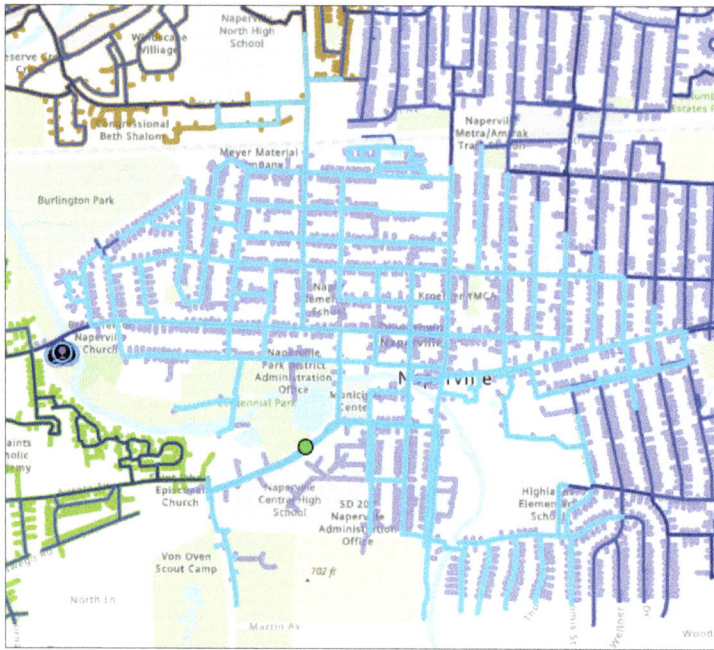

In this tutorial, you learned how to create a function barrier and saw how the parameters of the function affect trace results.

Tutorial 13-2: Understanding and creating filter function barriers

Filter function barriers provide another method to restrict your trace results using a numeric function, a network attribute, and a logical operator, in addition to function barriers. Filter function barriers work slightly differently from function barriers.

Filter function barriers can be useful when running a trace where the starting point is far from the subnetwork controller because the filter function barrier won't stop the trace from running, even if the distance to the subnetwork controller exceeds the function barrier parameters.

This distinction can be a challenging to understand, but the easiest way to see the difference in practice is to revisit the trace you ran in tutorial 13-1. You'll use the same starting point and trace parameters but decrease the **Value** field of the function barrier.

Decrease the Value field and rerun the trace

1. In the **Geoprocessing** pane, apply the same **Trace** tool parameters from tutorial 13-1.

2. Expand the **Function Barriers** section and change **Value** to 4000.

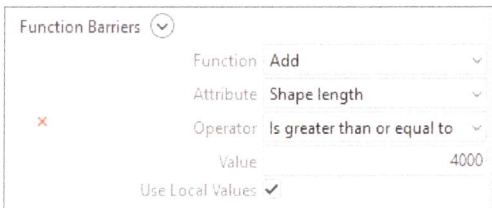

3. Click **Run**.

   The trace will fail and generate the following error:

   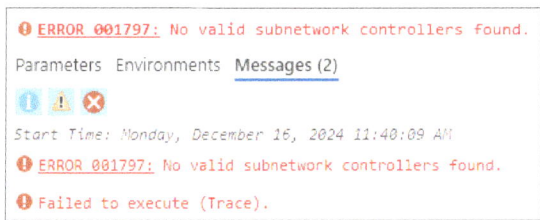

   The reason the trace failed is that the distance from the starting point to the subnetwork controller exceeded 4,000 feet. Because the function barrier restricted all traversal (including back to the subnetwork controller greater than 4,000 feet), the trace couldn't finalize because it couldn't locate the required subnetwork controllers.

   Cases such as this are examples of when filter function barriers can be helpful; regardless of the distance of the starting point from the subnetwork controller, they will succeed and return trace results. This occurs because the filter function barriers are considered after the traversability barriers (such as condition barriers or function barriers).

   Now that you understand more about the difference between function barriers and filter function barriers, you'll create a trace using a filter function barrier.

Set parameters for a filter function barrier

4. In the **Locate** pane, search for Wtr-Mn-1330.

5. Right-click the result, click **Add To Selection**, and click **Zoom To**.

6. Place a **starting point** at the midpoint of the **Wtr-Mn-1330** line.

7. In the **Geoprocessing** pane, use the same starting point and downstream trace parameters from tutorial 13-1 for the **Water Domain**, **Target Tier**, and **Tier** fields.

8. In the **Function Barriers** section, click the **Remove** button (red *X*).

9. Expand the **Filters** group. In the **Filter Function Barrier** section, apply the following parameters:

   - **Function**: Add
   - **Attribute**: Shape length
   - **Operator**: Is greater than or equal to
   - **Value**: 100

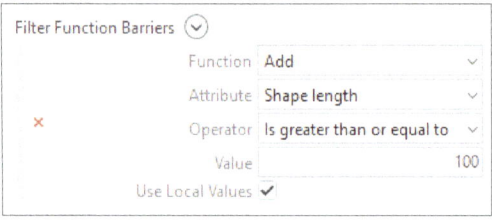

10. Click **Run**.

    The trace succeeds and returns a selection set of features within 100 feet downstream of the starting point.

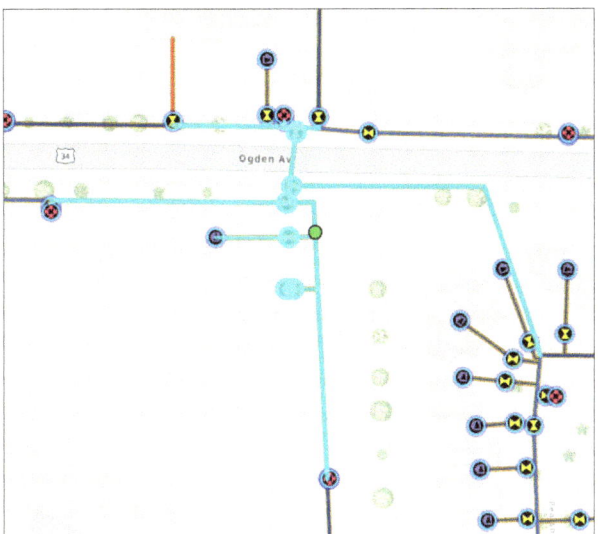

In this instance, the starting point is more than 100 feet from the subnetwork controller, but because you used a filter function barrier, the trace was able to locate the subnetwork controllers and return the expected trace results 100 feet downstream of the starting point.

In this tutorial, you learned about the difference between function barriers and filter function barriers—including when to use each—and you used a filter function barrier within a trace.

On your own

Create a list of function barrier conditions to use in your utility network.

Take the next step

Explore using different function barriers or filter function barriers such as **Subtract** and **Count** within other directional traces and study how the trace performs.

Summary

In this chapter, you learned more about function barriers and filter function barriers and used them within a trace.

Workflow

1. Place a starting point.
2. Perform a directional trace with Add Function Barrier (tutorial 13-1) or Add Filter Function Barrier (tutorial 13-2) configured to trace the shape length of 8,000 feet (tutorial 13-1) or 100 feet (tutorial 13-2).
3. Study the trace results.

CHAPTER 14
Using condition and filter barriers

Objectives

- Create condition barriers and examine how they affect trace results.
- Create filter barriers and examine how they affect trace results.

Introduction

In this chapter, you'll learn about condition barriers and filter barriers.

Filter barriers can be helpful when it may be difficult for a directional or subnetwork trace to locate the subnetwork controller because of condition barriers. Filter barriers are evaluated after condition barriers when running a trace.

Condition barriers limit the path a trace can take by using network attributes and categories within the utility network to help determine whether a trace should proceed. Condition barriers can be added to the utility network using the **Set Subnetwork Definition** geoprocessing tool or dynamically added to any subnetwork trace using the **Trace** geoprocessing tool.

Adding condition barriers to the trace configuration properties using the **Set Subnetwork Definition** tool allows parameters to be prepopulated in the Trace pane when running a subnetwork trace.

Tip: For information about how to set up your utility network project, see the "Getting Started" section. This chapter features the Water Distribution Utility Network Foundation datasets.

Tutorial 14-1: Creating and accessing a condition barrier

In this tutorial, you'll learn how to access condition barriers within trace configuration properties, examine how condition barriers are built into the utility network, and add a condition barrier to a subnetwork trace.

Review utility network properties

1. From the **Datasets_For_UN_Skills_Book** folder, open **Water Distribution Utility Network Foundation.aprx**.

2. In the **Water Distribution Editor** map, browse to the **Contents** pane.

3. Right-click the **Water Utility Network** layer and click **Properties**.

4. In the **Layer Properties** window, click the **Network Properties** tab.

5. Expand the **Water Network** section and expand the **Tiers** table.

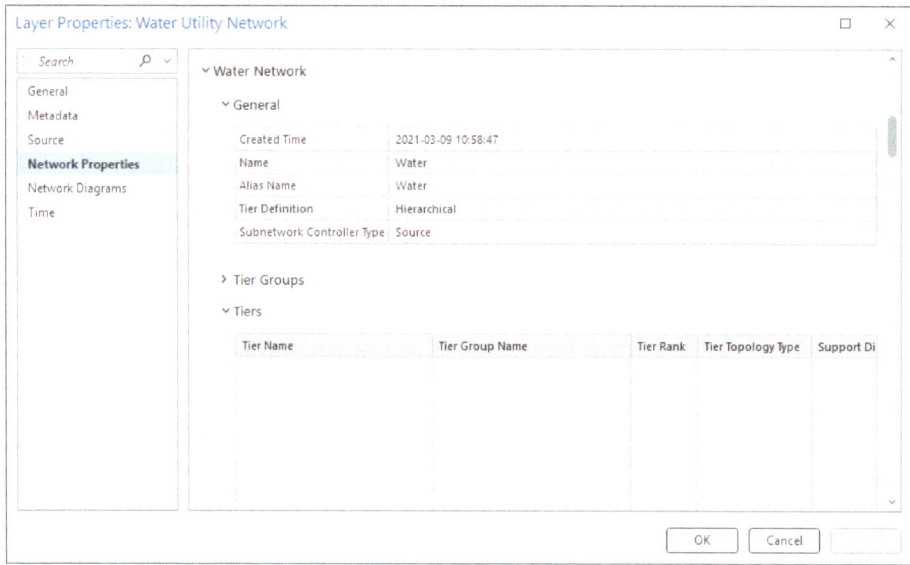

6. In the **Tiers** table, scroll over to the **Trace Configuration** column, which is near the end of the table. Examine the properties for the **Water System** row.

```
Include Containers: True
Include Content: True
Include Structures: True
Validate Locatability: False

Condition Barriers:
    P:Device Status IS_EQUAL_TO SPECIFIC_VALUE Closed OR
    Lifecycle Status DOES_NOT_INCLUDE_ANY SPECIFIC_VALUE In Service and To Be Retired OR
    Category IS_EQUAL_TO SPECIFIC_VALUE CP Only

Apply Traversability To: BOTH_JUNCTIONS_AND_EDGES

INCLUDE_BARRIERS

Summaries:
    ADD Measured Length Line Asset Group IS_EQUAL_TO SPECIFIC_VALUE 1 subnetworklength
    ADD Measured Length Line Asset Type IS_EQUAL_TO SPECIFIC_VALUE 1 transsubnetworklength
    ADD Measured Length Line Asset Type IS_EQUAL_TO SPECIFIC_VALUE 2 distsubnetworklength
    COUNT Device Asset Group Category IS_EQUAL_TO SPECIFIC_VALUE Disconnecting servlvcnt
    COUNT Device Asset Group Category IS_EQUAL_TO SPECIFIC_VALUE Protective protcnt
    COUNT Device Asset Group Device Asset Group IS_EQUAL_TO SPECIFIC_VALUE 7 hydrantcnt
    COUNT Device Asset Group Device Asset Group IS_EQUAL_TO SPECIFIC_VALUE 12 serviceconcnt
```

7. Close the **Layer Properties** window.

> **Tip:** Condition barriers within the subnetwork definition can be assigned to each tier. Condition barriers that are part of the subnetwork definition, as shown in the preceding image, can be removed from the Trace pane manually if not needed for the specific trace being run.

Next, you'll look at a few condition barriers and explore the role they play in a subnetwork trace.

Run a trace with preconfigured parameters

8. On the ribbon, click the **Utility Network** tab. In the **Tools** group, click the **Subnetwork** tool.

   The **Trace** geoprocessing tool opens, preconfigured for a subnetwork trace.

9. In the **Geoprocessing** pane, set the following parameters and keep the default values for all other parameters:

   - **Domain Network**: Water
   - **Tier**: Water System
   - **Subnetwork**: Naperville Distribution

10. Click **Run**.

    The trace will return 55,328 features within the Naperville distribution subnetwork.

    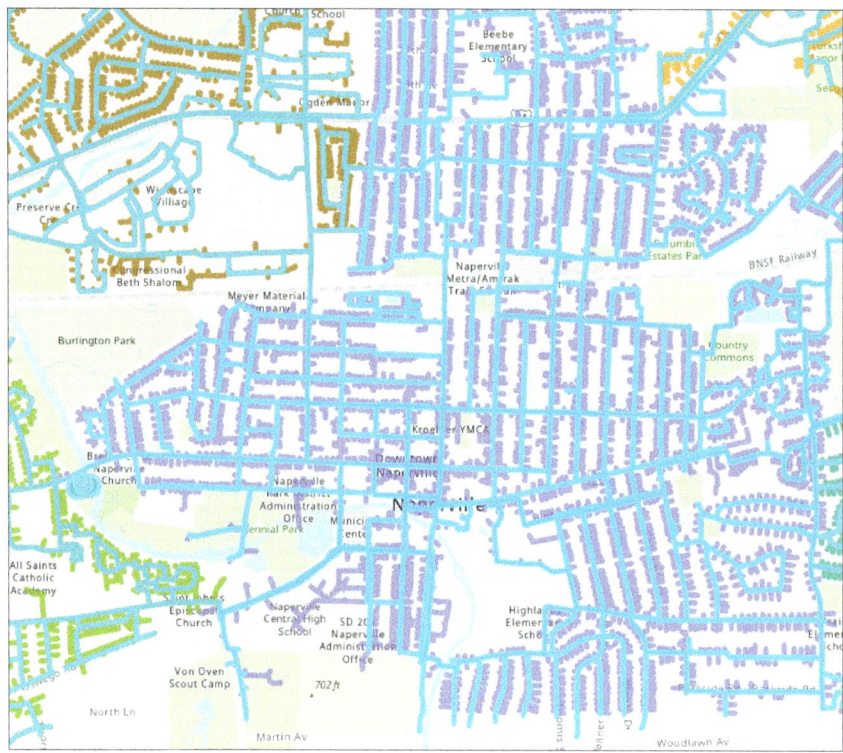

11. Scroll down in the **Trace** tool pane and expand the **Traversability** section.

12. Under **Condition Barriers**, click **Add another**.

    You'll add a condition barrier to have water device features with an asset type of **Supply** acting as condition barriers.

    > **Tip:** Logical operators (AND, OR) can be used for multiple condition barriers together in a single trace.

13. In the **Geoprocessing** pane, set the following parameters for the new condition barrier:

    - **Combine Using** (in the previous condition barrier block): Or
    - **Name**: Device Asset Type

- **Operator**: Is equal to
- **Type**: Specific value
- **Value**: 170

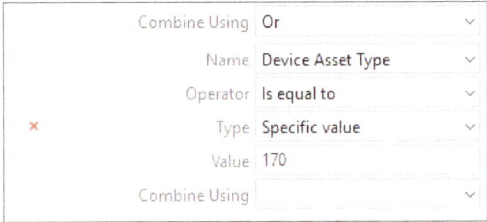

Note: The Value parameter expected by the system is the code, not the description, for the domain or subtype used by the field contained within the network attribute. In this example, the code for the asset type is 170, but the description is Supply.

14. Click **Run**.

The following image shows the results of the Naperville distribution subnetwork trace using only the preconfigured condition barriers within the subnetwork definition.

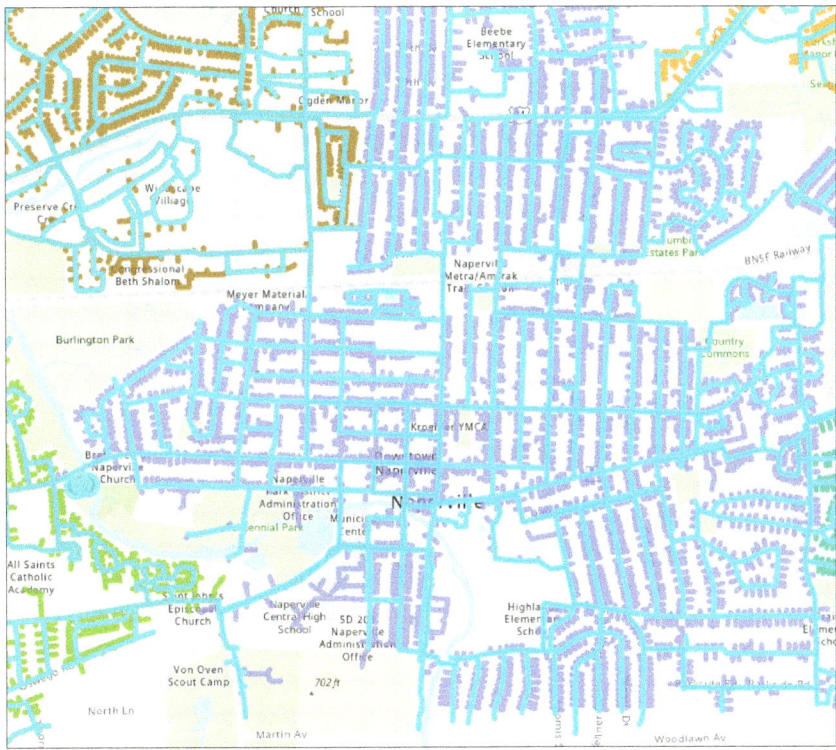

The following image shows the same portion of the subnetwork with the condition barrier you added previously.

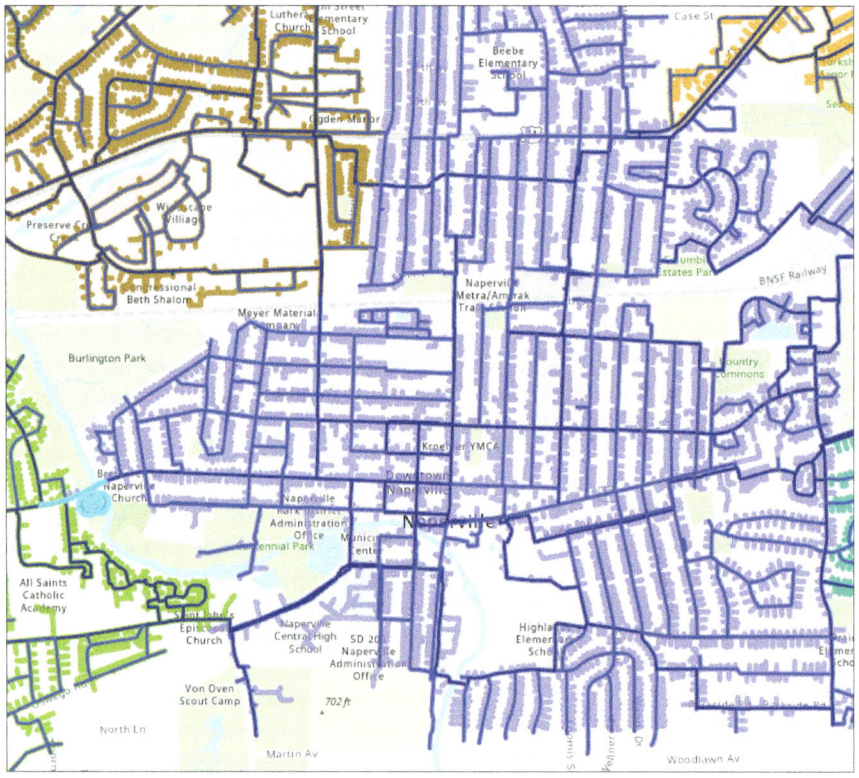

The rest of the subnetwork has been excluded from trace results because of the new condition barrier that's in place for devices with an asset type of **Supply**. The trace found a device that met this condition and halted the trace, returning only the results it found before encountering the device that met the condition barrier.

Next, you'll build on your skills to create filter barriers.

Tutorial 14-2: Creating a filter barrier

In this tutorial, you'll learn how to create filter barriers using the **Water Distribution Utility Network Foundation**. Filter barriers can be used with condition barriers, which you learned about in tutorial 14-1, to stop a trace if a condition is met.

You'll run an isolation trace of the water network to determine what devices you need to close if there's a leak.

Configure and run an isolation trace

1. Open the **Locate** pane and search for a line with an **asset ID** of Wtr-Mn-1442.

2. Right-click the result, click **Add To Selection**, and click **Zoom To**.

3. In the **Trace** pane, place a starting point on the midpoint of the **Wtr-Mn-1442** line.

4. Click the **Utility Network** tab. In the **Tools** group, click the **Isolation** tool.

    This tool will configure an isolation trace.

    > *Note: Isolation traces require the use of a filter barrier. The trace will fail unless a filter barrier is provided.*

5. In the **Isolation** tool pane, set the following parameters, but keep the default values for all other parameters:

    - **Domain Network**: Water
    - **Tier**: Water System

6. Expand the **Filter** section, and under **Filter Barriers**, set the following parameters:

    - **Name**: Category
    - **Operator**: Is equal to
    - **Type**: Specific value
    - **Value**: Isolating

    > **Tip:** The filter barrier is using a network category called Isolating in the utility network. As a result, features from multiple asset groups and asset types will serve as barriers when running this trace even though only this single network category is referenced. Using network categories in this fashion allows you to configure barriers in a utility network.

7. Run the trace and review the results.

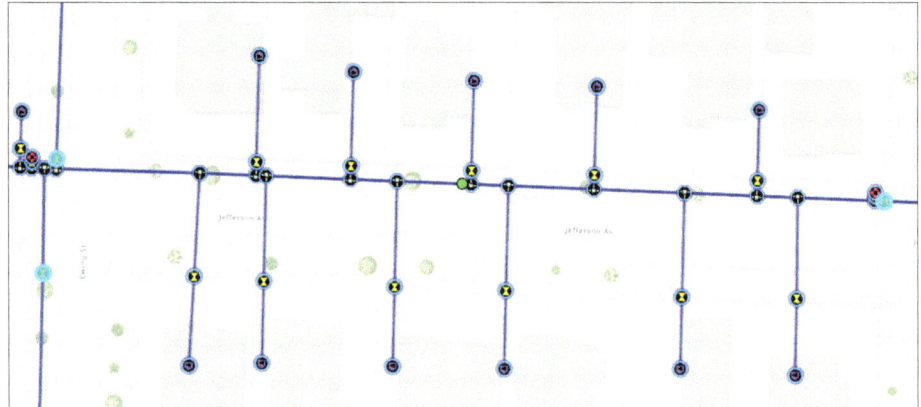

This trace returned a selection of all the valve features in the **Isolating** category that could be used to seal off this portion of the water network.

This procedure is an example of how the configurability of the utility network model can be combined with the trace capabilities to help answer real-world questions that ArcGIS users ask every day.

On your own

Study your organization's data to determine what parts of the infrastructure can be used as condition barriers or filter barriers.

Take the next step

Create custom network categories or network attributes to use as condition or filter barriers in different trace types. Analyze how barriers restrict flow within the network to understand where additional barriers may be needed as you develop traces specific to your workflows.

Summary

In this chapter, you learned about the difference between condition barriers and filter barriers, how to create both types of barriers, and learned how they affect trace results through running a subnetwork trace and an isolation trace.

Workflow

1. Create a starting point.
2. Create barriers.
3. Run a subnetwork trace with a condition barrier applied.
4. Run an isolation trace with a filter barrier applied.
5. Study the trace results.

CHAPTER 15
Working with functions

Objectives

- Learn how to create functions.
- Learn how functions help answer real-world questions.

Introduction

In this chapter, you'll learn how to create a function to determine the number of devices downstream of a starting point in the network.

Functions within the trace framework allow you to understand what your trace returns by performing operations on the features returned in the output. Functions can be used with any trace type in the utility network, and multiple functions can be run simultaneously within a single trace.

For example, functions can be used to count the number of abandoned devices within a subnetwork or add the measured length for all lines found in a downstream trace.

Chapter 15: Working with functions

Tip: For information about how to set up your utility network project, see the "Getting Started" section. This chapter features the Electric Utility Network Foundation datasets.

Tutorial 15-1: Creating a function

In this tutorial, you'll learn how to create a function using network attributes to locate specific features within the network. You'll also learn how to access function results in ArcGIS Pro.

Create the starting point

1. From the **Datasets_For_UN_Skills_Book** folder, open **ElectricUtilityNetworkFoundation.aprx**.

2. In the **Electric Network Editor** map, open the **Locate** pane.

3. In the **Locate** pane, search for a device with an **asset ID** of MV-COND-496. Right-click the result and click **Add to Selection** and **Zoom To**.

This overhead conductor feature is where the function will be calculated when running the trace.

4. In the **Trace** pane, place a starting point on the **MV-COND-496** line.

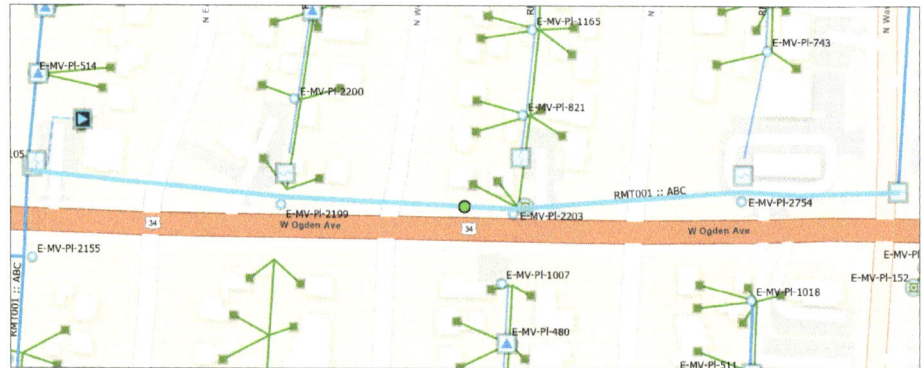

Configure the downstream trace

5. On the ribbon, click the **Utility Network** tab. In the **Tools** group, click the **Downstream** tool to open the **Trace** geoprocessing tool.

6. In the **Trace** tool pane, apply the following parameters but keep the default values for all other parameters:

   - **Domain Network: Electric**
   - **Tier**: Electric Distribution
   - **Target Tier**: Electric Secondary

   Next, you'll create a function.

7. Scroll down and expand the **Functions** section.

   In the **Functions** section, you'll see several parameters: **Function**, **Attribute**, **Filter Name**, **Filter Operator**, **Filter Type**, and **Filter Value**.

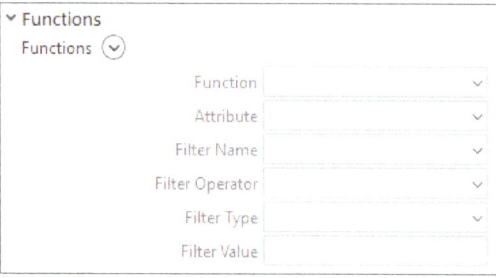

You want to understand how many devices in the **Electric Device** table downstream of the starting point on the **MV-COND-496** line can be operated on multiple phases (a condition known as phase operable). You need to use the power of tracing in combination with a network attribute within the function.

Before creating the function parameters, you need to examine the **Assignments** table of the **Electric Utility Network** layer.

Examine the Assignments table

8. In the **Contents** pane, right-click the **Electric Utility Network** layer and click **Properties**.

9. Click the **Layer Properties** tab and scroll down to expand the **Attributes and Assignments** section. Browse to the **Assignments** table.

Your question pertains to how the device operates, so you need to look for the **E:Field Operation** network attribute.

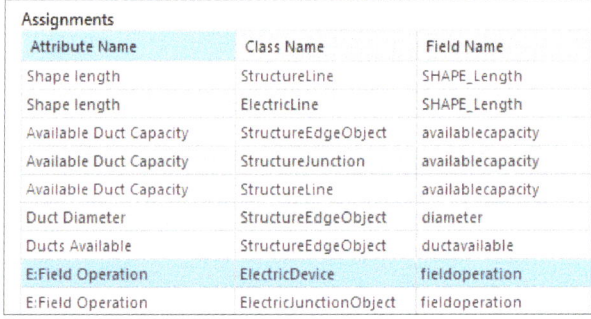

Tip: The Assignments table is a great resource to see which fields in feature classes are assigned to network attributes.

You want to locate only those devices that are phase operable, so you'll set the following parameters.

Set the parameters

10. In the **Geoprocessing** pane, under **Functions**, apply the following parameters to your function:

    - **Function**: Count
    - **Attribute**: E:Field Operation
    - **Filter Name**: E:Field Operation
    - **Filter Operator**: Is equal to
    - **Filter Type**: Specific value
    - **Filter Value**: Phase Operable

    Tip: Filter Name is an optional parameter.

11. Click **Run**.

12. When the trace is complete, click **View Details** to open the **View Details** window.

13. In the **View Details** window, scroll to the right to see the **Result** column.

    You can see that 35 results have been returned in the function output. This number represents the devices for which the **E:fieldoperation** value in the **Electric Device** feature class is set to phase operable in the downstream trace results. The **Filter Value** of **2** corresponds to phase operable.

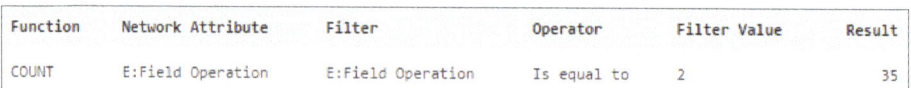

In this tutorial, you learned the basics of a function and how network attribute configuration uses utility network properties. You also learned how a network attribute can be used in a function to provide more information about the state of the utility network. Finally, you learned how to view and interpret function results.

On your own

Try out traces with multiple functions.

Take the next step

Read about other function operators such as **Min**, **Max**, and **Average** and test them within traces.

Summary

In this chapter, you learned how to configure functions and explored the role they play in tracing in the utility network.

Workflow

1. Place a starting point.
2. Configure a trace with a function.
3. Run the trace.
4. Evaluate function output.

CHAPTER 16
Working with trace output configurations

Objectives

- Explore different trace output configurations.
- Configure different trace outputs.

Introduction

You've learned in previous chapters that tracing in the utility network is a powerful tool that can answer critical questions about the network's state and configuration. This chapter explores how you can use these results to visualize the data in new ways or share it with others in your organization.

Trace output configurations provide options for changing how your trace results are returned and how they can be visualized or exported to external files. In this chapter, you'll learn about three types of trace output configurations: output asset types, output conditions, and result types. You'll also explore the types of information that a trace can return. Last, you'll learn how to save customized trace templates for reuse.

Tip: For information about how to set up your utility network project, see the "Getting Started" section. This chapter features the Electric Utility Network Foundation datasets.

Tutorial 16-1: Using output asset types

In this tutorial, you'll learn about output asset types and how this functionality is used to return features in one or more asset types in a selection set of trace results. This capability is useful when you need to trace through the network but are only concerned with a small set of the results that your trace will return. You'll use output asset types to find all poles downstream of a starting point.

Add a starting point

1. From the **Datasets_For_UN_Skills_Book** folder, open **ElectricUtilityNetworkFoundation.aprx**.

2. In the **Electric Network Editor** map, open the **Locate** pane.

3. In the **Locate** pane, click the **Layer Search** tab and search for a line with an **asset ID** of MV-COND-123. Right-click the result and click **Add To Selection**. Click **Zoom To**.

4. In the **Trace** pane, add a starting point on the **MV-COND-123** line.

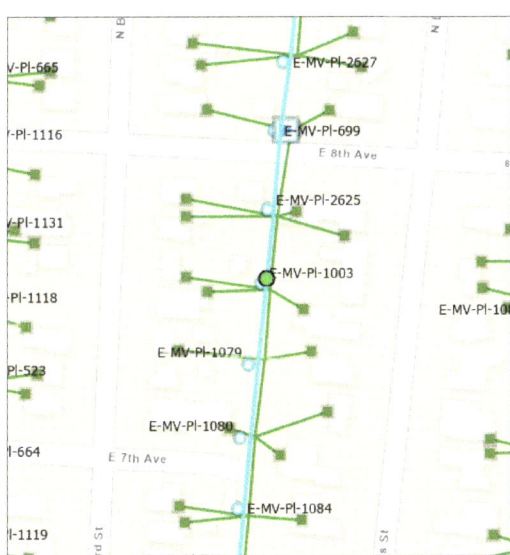

Next, you'll configure the downstream trace.

Configure the downstream trace

5. On the ribbon, click the **Utility Network** tab. In the **Tools** group, click the **Downstream** tool to open the **Trace** geoprocessing tool.

6. In the **Trace** tool pane, apply the following parameters:

    - **Domain Network**: Electric
    - **Tier**: Electric Distribution
    - **Target Tier**: Electric Secondary

7. Keep the default values for all other parameters. Scroll down to expand the **Output** section.

8. For **Output Asset Types**, click the drop-down list and select **StructureJunction/Electric Medium Voltage Pole/Pole MV**.

    > **Tip:** The value in the list is a combination of the asset group feature class and asset type. The name of the feature class is Structure Junction, the name of the asset group is Electric Medium Voltage Pole, and the name of the asset type is Pole MV.

9. Click **Run**.

10. When the trace finishes, in the bottom-right corner of the map, click **Selected Features: 9**.

    The map extent shows all the poles found by your trace.

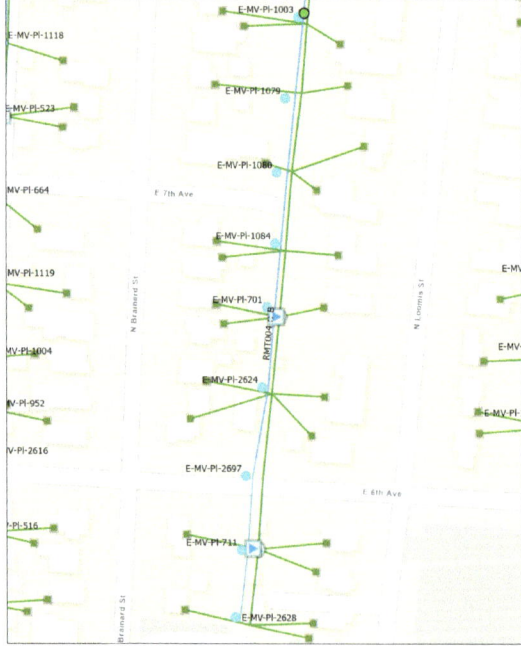

In this tutorial, you learned how to use output asset types in trace output configurations.

Tutorial 16-2: Using output conditions

Next, you'll learn more about how to use output conditions to return features that meet certain criteria when running a trace.

Output conditions allow you to select which network attributes or categories will be returned in a trace result. Unlike tutorial 16-1, in which the trace returned features with a specific asset type, output conditions use network attributes or categories and logical conditions to provide only a specified set of features for which the condition is met.

For example, you can use output conditions to return only those features that have a certain network attribute or network category value.

In this tutorial, you'll use the downstream trace from tutorial 16-1 to return only those features that have the network attribute **E:Phases Normal** and that also have an attribute value of **II or B, or b**.

Configure the downstream trace

1. Return to the downstream trace configuration from tutorial 16-1. In the **Trace** pane, confirm that **Trace Type** is set to **Downstream**.

2. Confirm that **Starting Point** is still present on the **MV-COND-123** line.

3. Scroll to the bottom of the pane. In the **Output** section, click the **Remove** button (red *X*) to remove the parameters for **Output Asset Types**, leaving them blank.

4. For **Output Conditions**, apply the following parameters:

   - **Name**: E:Phases Normal
   - **Operator**: Is equal to
   - **Type**: Specific value
   - **Value**: II or B, or b

Configuring the output conditions this way will return only those devices that use the **E:Phases Normal** network attribute and where the **phasesnormal** field has a value of **B**.

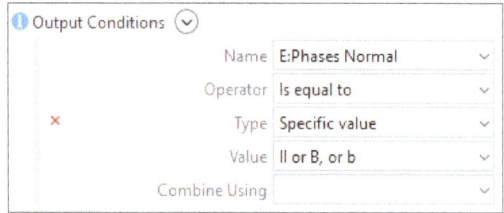

Tip: The E:Phases Normal network attribute is applied to multiple feature classes across the domain network, as can be seen in the utility network properties on the Attributes and Assignments tab. By querying a network attribute for a specific value instead of a single field on a feature class, the trace can return results from these five feature classes in instances where the condition specified is met.

E:Phases Normal	ElectricDevice	phasesnormal
E:Phases Normal	ElectricEdgeObject	phasesnormal
E:Phases Normal	ElectricJunction	phasesnormal
E:Phases Normal	ElectricJunctionObject	phasesnormal
E:Phases Normal	ElectricLine	phasesnormal

Run the tool

5. Run the **Trace** tool with the output conditions applied and examine the map extent.

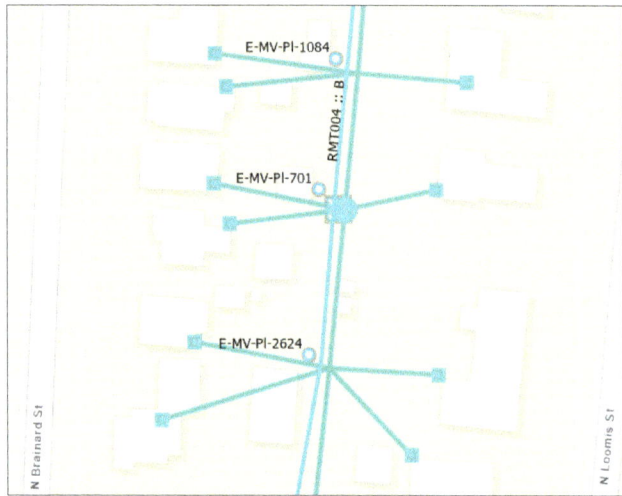

The trace returned only 61 features because there are 61 features downstream of the starting point with the **E:Phases Normal** network attribute value of **B**.

6. Remove the output condition parameters and rerun the **Trace** tool.

   The trace returns 72 features because it includes all devices downstream of the starting point, regardless of the value in the **phasesnormal** field. No output condition has been applied, so there's no restriction on the output trace results.

   In this tutorial, you learned how to create output conditions and saw how they affect trace results.

Tutorial 16-3: Using result types

Next, you'll explore various result types in the utility network.

Result types are the third output parameter you can configure when running a trace. Through the many traces you have run, you have become familiar with the default result type, a selection set of features in the map. But other result types, including aggregated geometry and JSON file outputs, contain trace results.

When aggregated geometry is selected as the result type, the trace outputs will return three feature classes—a point, line, and polygon feature class—that combine all geometries of each type in the trace result into a single feature class for each geometry type.

Result types can be helpful when you want to quickly visualize the geographic extent of your trace or create a table of all the points or polygons returned by the trace for further analysis. Result types can also be configured to produce JavaScript Object Notation (JSON) file outputs of the connectivity, elements, features, and containment and attachment associations returned by a trace. Some organizations use this capability to facilitate integration of utility network data with other business systems. Multiple result types can be returned in a single trace for flexibility in how you work with your trace results.

In this tutorial, you'll use the downstream trace you ran in tutorials 16-1 and 16-2 but add a different result types.

Set result types and examine the results

1. Return to the downstream trace configuration from tutorial 16-1 and tutorial 16-2. Clear any output conditions or configurations from the previous tutorials.

2. Scroll down to the **Output** section to find the **Result Types** parameter.

3. Set the **Result Types** parameters to **Selection**, **Connectivity**, **Features**, and **Aggregated Geometry**.

By setting different result types in the same trace, you'll increase the flexibility in the utility network to tailor trace results to meet your needs.

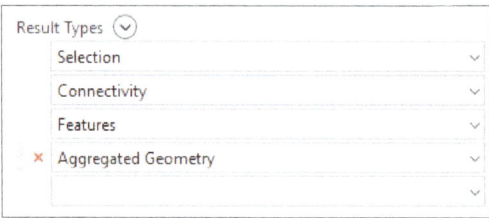

4. Examine the new parameters in **Result Types**.

> **Tip:** The additional configuration options aren't visible by default and only appear as result types that require them to be added.

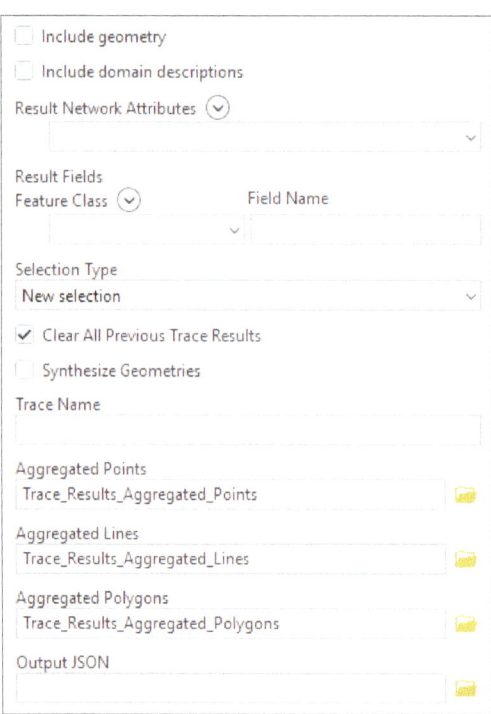

Before you use these configured result types, it's important to understand how these options work. This information is essential to understand how the trace will perform and what the output will (or won't) contain.

The **Include geometry** check box refers to whether spatial geometry information should be included in the output JSON you'll name at the bottom of the pane. This information may be helpful for other external business systems to illustrate the spatial location and characteristics of data.

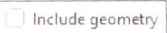

The **Include domain descriptions** check box allows you to change how data is represented in the output JSON when using result fields. By default, the field values in the JSON with a domain use the domain code—for example, 7—instead of the domain description of Transformer. Employing this parameter makes the output JSON easier to understand.

Result network attributes allow you to add network attributes on features returned by the trace to the output JSON if using result types: features, elements, containment and attachment, or connectivity. Multiple network attributes can be exported as part of a single trace.

Result fields allow you to add fields within specific feature classes. The value of the field will be written to the output JSON if the feature is included in the trace results. Multiple fields from different feature classes can be included. For example, a result field can be added to include the creation date of a feature to the output JSON. This capability can provide other external business systems with data from the utility network.

For the **Selection Type** field, the **New selection** option provides flexibility in how the new selection set of features interacts with an existing selection set in the map. For example, if features are currently selected, the features returned by the trace can be appended onto that selection set.

The **Clear All Previous Trace Results** check box allows you to append new aggregated geometry results. This parameter is enabled by default.

The **Synthesize Geometries** check box allows you to create geometry for associations and edge objects. This parameter is disabled by default.

Trace Name allows for a unique name to be provided when the **Aggregated Geometry** setting is selected as a result type. The name you assign will appear in the Contents pane and help distinguish the trace results from other feature classes and data in the map.

Aggregated Points, **Aggregated Lines**, and **Aggregated Polygons** allow you to name the feature classes that are created as part of the aggregated geometry result type output and specify the feature classes to which the results are written.

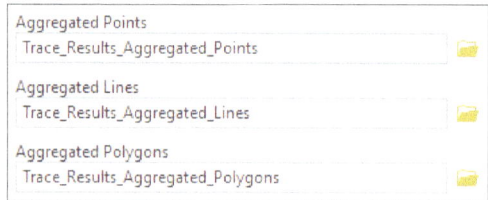

Output JSON is where the name of the JSON is provided, if the result type is specified as features, elements, containment and attachment associations, or connectivity. ArcGIS Pro will prompt you to select a location on the computer to save the output JSON.

This section provided a useful review of the various result types and configuration parameters. Next, you'll explore trace configurations.

Add a trace configuration

As you close out your journey through tracing within the utility network, there is one final topic to cover: storing your configured traces.

Creating the parameters for a complicated trace can be time-consuming and tedious, but it doesn't need to be. The utility network allows you to save and reuse a set of trace parameters when you need them. The file that holds all the information about a trace is called a trace configuration. Once the trace configuration is imported into the **Trace** tool pane, all parameters stored in the file will show up in the pane and the trace will be ready to run.

To create a trace configuration, use the **Add Trace Configuration** geoprocessing tool to create a set of trace parameters and then save them for reuse.

5. In the **Geoprocessing** pane, search for and open Add Trace Configuration (Utility Network Tools).

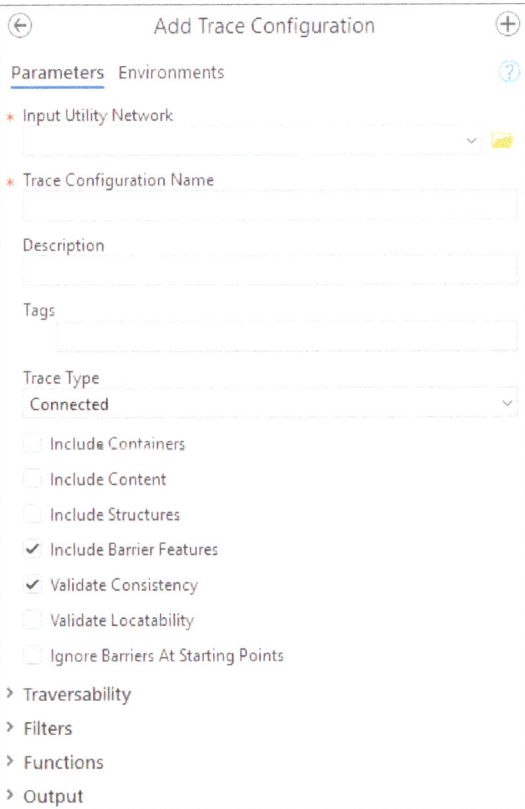

Once a trace configuration exists in the utility network, click the **Use Trace Configuration** box at the top of the **Trace** tool and you'll be prompted to load the trace configuration.

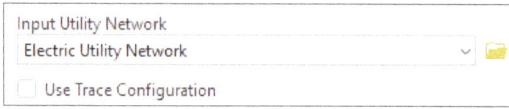

In this chapter, you learned about different types of trace outputs and trace configurations. Happy tracing!

On your own

Experiment with result types to visualize your trace results in different ways and create JSON files of your trace outputs with various fields and network attributes.

Take the next step

Create trace configurations for the most common traces that your organization may use to save time.

Summary

In this chapter, you learned the differences between a variety of trace output types, how to customize trace output types with different parameters, and the benefits of using trace configurations to reuse commonly run traces.

Workflow

1. Configure a downstream trace.
2. Add different output types to the trace:
   - Output configurations
   - Output conditions
   - Result types
3. Run the trace.

CHAPTER 17
Working with network diagrams

Objectives

- Select a subnetwork using the Find Subnetwork layer.
- Create a network diagram from a selected isolation subnetwork.
- Demonstrate feature selection propagation.

Introduction

Network diagrams provide a unique insight into utility network features. Engineers frequently use schematics and diagrams to highlight critical features and their relationships with one another in circuits and feeders. Starting with a set of selected features, users can create general map diagrams of those selected features. Diagrams create a visualization that enables users to show the link across key features or critical protective devices.

> **Tip:** For information about how to set up your utility network project, see the "Getting Started" section. This chapter features the Gas and Pipeline Utility Network Foundation dataset.

Tutorial 17-1: Creating a network diagram from a traced subnetwork

Most diagram outputs represent network circuits or feeders. This tutorial will demonstrate the process to generate a network diagram from selected features.

Select a subnetwork to generate a network diagram

1. From the **Datasets_For_UN_Skills_Book** folder, open **Gas and Pipeline Utility Network Foundation.aprx**.

2. In the **Gas and Pipeline Network Editor** map, activate the **Utility Network** tab.

3. In the **Subnetwork** group, click the **Find** button to open the **Find Subnetworks** pane.

   In the **Find Subnetworks** pane, the table lists available subnetworks that are visible within the current map extent.

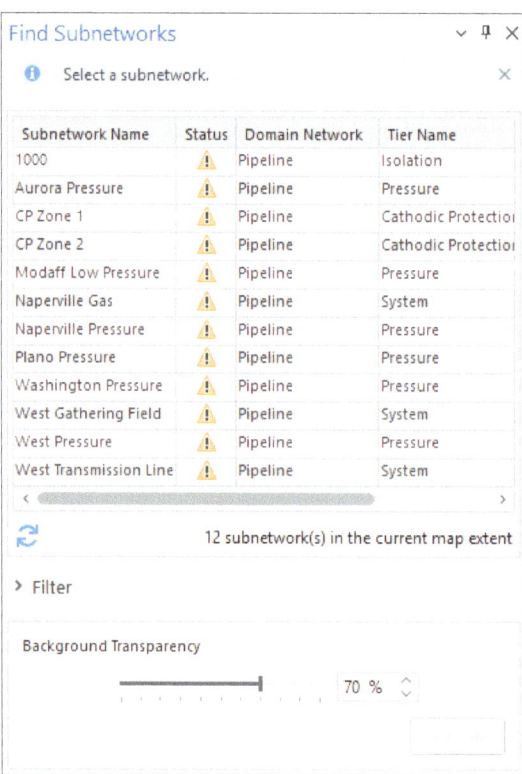

Trace an isolation network

The first record is **Isolation Zone 1000**.

4. Right-click the first record and click **Zoom To**.

5. Right-click again and click **Trace Subnetwork**.

   The map zooms to **Isolation Zone 1000** and traces it.

Create a network diagram from selected features

6. On the ribbon, click the **Utility Network** tab, and in the **Diagram** group, click the **New** button.

   The **Diagram** window pops up, displaying all features configured in the basic layout.

Tip: Subnetworks can be significantly different between electric or communications networks compared with pipe-based systems, such as gas or water.

Select features from the diagram space to propagate the selection to the map

7. On the ribbon, on the **Network Diagram** tab, locate the **Selection** group and click the **Select** tool.

8. In the diagram, use the **Select** tool to drag a box over any collection of features.

   This example selects a collection of features in the top-right corner of the subnetwork.

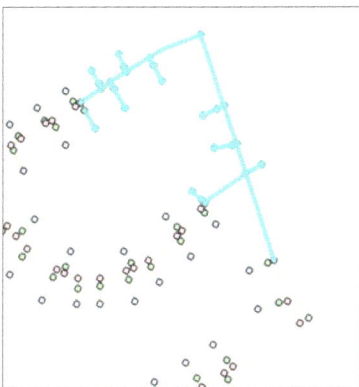

9. On the **Network Diagram** tab, in the **Selection** group, click the **Apply to Map** button.

10. Position the **Diagram** window beside the **Gas and Pipeline Network Editor** map window.

You can see the results of the selection propagation on the **Gas and Pipeline Network Editor** map.

> **On your own**
>
> Use these steps to build network diagrams in water or electric foundation networks.

Take the next step

Network diagrams can be edited and modified much like the map. Take the diagram and edit lines and points to see how you can edit a diagram for clarity.

Summary

In this chapter, you learned how to create a basic network diagram from utility network features, select elements in the diagram, and push that selection back to the map space.

Workflow

1. Browse to Isolation Subnetwork 1000 and run a trace using the Find Subnetworks pane.
2. Create a new diagram from the tool on the Utility Network tab.
3. Select features from the new diagram.
4. Propagate those features back to the map.

CHAPTER 18
Modifying diagrams with layout configurations

Objectives

- Use preconfigured network diagram layouts.
- Edit network diagrams.
- Use basic editing in network diagrams.
- Store and find stored diagrams.

Introduction

In this chapter, you'll learn how to work with layouts to customize the appearance of your network diagram using the **Gas and Pipeline Utility Network Foundation** dataset. Specifically, you'll learn how to build a smart tree layout configuration to customize the look of your network diagram.

> **Tip:** For information about how to set up your utility network project, see the "Getting Started" section. This chapter features the Gas and Pipeline Utility Network Foundation dataset.

Tutorial 18-1: Applying a layout to a network diagram

In this tutorial, you'll use the same network diagram that you used in chapter 17. First, you'll open **Isolation Subnetwork 1000** using the **Find Subnetworks** pane.

Open the isolation subnetwork

1. From the **Datasets_For_UN_Skills_Book** folder, open **Gas and Pipeline Utility Network Foundation**.aprx.

2. In the **Gas and Pipeline Network Editor** map, open the **Find Subnetworks** pane.

3. Zoom to and run a trace on **Isolation Subnetwork 1000** to select the network features.

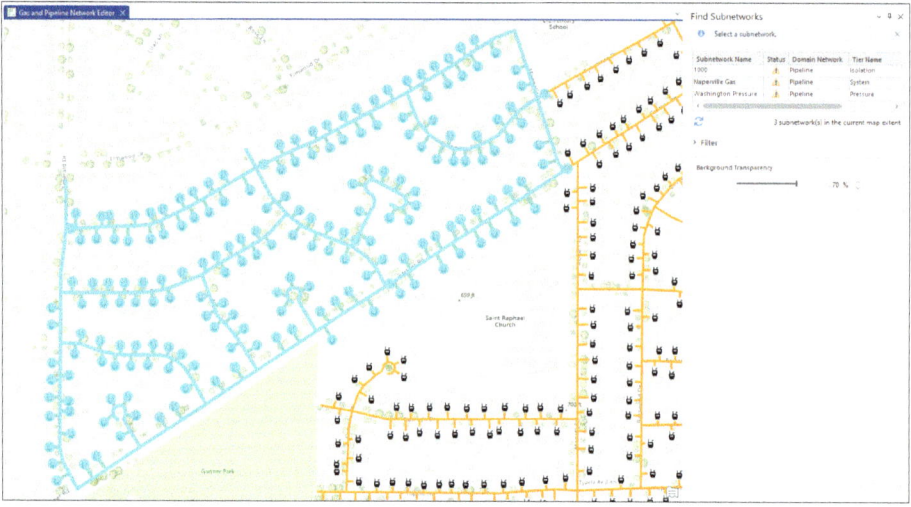

Create a network diagram

4. On the **Utility Network** tab, in the **Diagram** group, click **New** to create a new diagram.

   The **Diagram** window opens with a network diagram.

Configure the diagram layout

5. On the **Network Utility** tab, in the **Layout** group, expand the **Diagram Layouts** menu.

6. In the **Smart Tree** section, select **Left to Right**.

   The smart tree layout modifies the network diagram based on the network properties of the selected network features (from upstream to downstream). Smart tree diagram tools create schematics similar to single-line diagrams often used by utilities of all industries.

Edit the diagram layout

In some cases, additional layout modifications are needed to suit the engineering requirements. Previously, a smart tree layout was used to establish an initial diagram layout. You can also create a point that will have all features re-arrayed based on the new starting point of the network.

The **Set Flags** tool allows you to establish a point, or root junction, which should be reconfigured to be on the left side of the diagram.

7. In the Layout group, click the **Set Flags** tool.

8. Place a flag anywhere on a feature toward the middle of the diagram.

   You can see the new root junction point set as a green dot on the map.

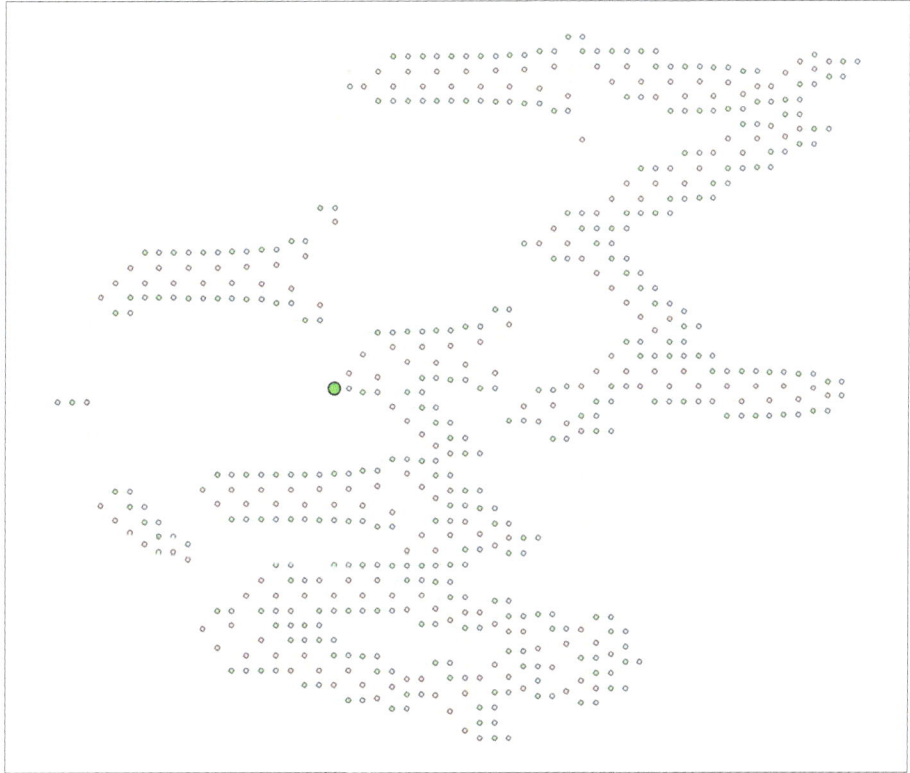

Place a root junction

The flag you set will force the layout to be reconfigured. Flags and root junctions are used to establish the start point for a diagram.

9. In the **Layout** group, select the **Smart Tree – Left to Right** layout again.

   The network diagram is reconfigured differently from the first layout earlier. The root junction that you placed is now the leftmost point of the diagram.

   Tip: Flags are optional parameters when generating diagrams.

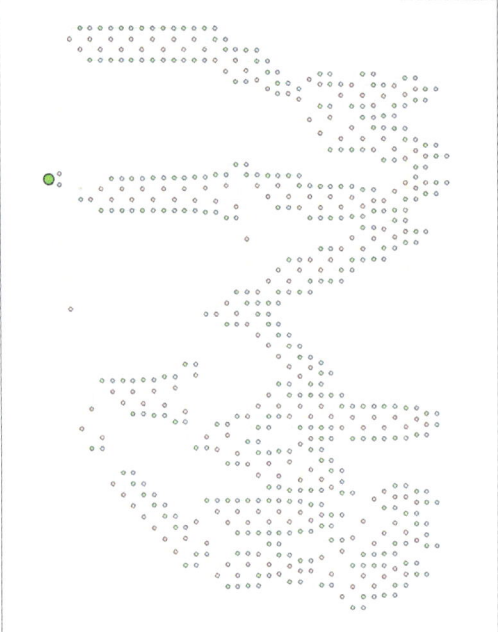

Edit network diagram features

10. Zoom to the left side of the diagram, where the root junction is.

11. On the ribbon, click the **Edit** tab. In the **Tools** group, click the **Edit Vertices** tool.

12. Modify the line by adding an angle (or any other modification).

13. In the **Manage Edits** group, click **Save** to save the new modification to the diagram.

Chapter 18: Modifying diagrams with layout configurations

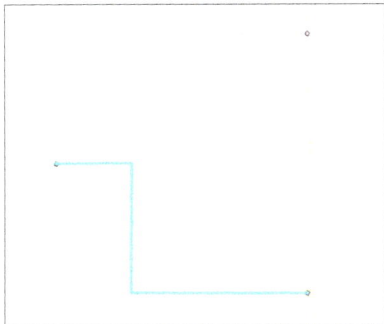

Tip: Modifying features using the Edit Vertices tool is similar to performing basic edits, which are covered in chapter 2 ("Basic Editing in ArcGIS Utility Network").

Store network diagrams

Now that the network diagram has been modified and edited, you'll store it.

14. On the ribbon, click the **Network Diagram** tab. In the **Manage** group, click **Store** to open the **Store Diagram** geoprocessing tool.

15. In the **Store Diagram** tool pane, apply the following parameters:

    • **Network Diagram Name:** Configured Network Diagram

16. Click **Run**.

    After it's been saved, the diagram can be retrieved in the **Find Diagrams** pane.

17. Click the **Gas and Pipeline Network Editor** map tab.

18. On the **Utility Network** tab, in the **Diagram** group, click the **Find** button to open the **Find Diagrams** pane.

    Tip: The Find Diagrams pane works just like the Find Subnetworks pane. They both populate the list based on the extent of the active map space.

On your own

Explore other foundation datasets to see the impact of various network diagram layout configurations.

Take the next step

Take a subset of your diagram and modify its layout. Network diagram layouts can be applied to the whole diagram space or a subsection of it.

Summary

In this chapter, you learned how to use trace results to generate a smart tree network diagram. You also learned how to add root junctions and edits to a diagram. Finally, you observed the process of storing diagrams.

Workflow

1. Create a network diagram.
2. Apply a diagram layout.
3. Add a root junction and apply the same diagram layout to see its impact on the diagram.
4. Store a network diagram.

CHAPTER 19
Configuring network controllers

Objectives

- Evaluate the subnetwork definition to determine valid subnetwork controllers.
- Configure a device as a subnetwork controller.

Introduction

Utilities of all types need to understand the flow of resources throughout their network to help their business, engineering, and operations teams make better decisions. One common requirement is a need to track where resources originate and where they go.

Within the utility network, subnetwork controllers play a critical role in establishing directionality to model the flow of resources. Specific features within the utility network can be classified as subnetwork controllers. After receiving this designation, those features can take on a special role and perform certain functions within the network. You have already seen the impacts of subnetwork controllers in previous chapters as you learned about directional tracing.

The utility network can have two types of subnetwork controllers: source-based and sink-based. When the source-based subnetwork controller is used, resources can only flow *away* from the device that's set as a controller. Circuit breakers in an electric network represent an asset that could be modeled in this fashion because electricity flows *away* from the circuit breaker. Contrastingly, in a utility network with sink-based subnetwork controllers, the resource can only flow *toward* the controller. A drain within a stormwater network represents an asset that could be modeled in this manner because water flows *toward* it. It's important to consider how resources move in the real world within your network and assign subnetwork controllers accordingly because all controllers on a given network must function as either sources or sinks.

As your utility network expands and changes through editing, the types of devices you use as subnetwork controllers and the number of controllers you need for each subnetwork may change. It's important to know how subnetwork controllers are configured and how to add one to the utility network. In this chapter, you'll learn more about subnetwork controllers, including how to create and configure them.

> **Tip:** For information about how to set up your utility network project, see the "Getting Started" section. This chapter features the Electric Utility Network Foundation datasets.

Tutorial 19-1: Evaluating the subnetwork controller definition and creating a controller

In this tutorial, you'll learn how to determine what devices can be used as subnetwork controllers and then assign a device to be a subnetwork controller.

1. From the **Datasets_For_UN_Skills_Book** folder, open **ElectricUtilityNetworkFoundation.aprx**.

2. In the **Electric Network Editor** map, browse to the **Contents** pane and right-click the **Electric Utility Network** layer. Click **Properties**.

3. In the **Layer Properties** window, click the **Network Properties** tab.

4. Scroll down and expand **Electric Network**. Expand **Tiers**.

5. Look for the column with the header **Valid Subnetwork Controllers**.

6. Scroll down to find the **Electric Distribution Primary** tier and review its asset groups and asset types under **Valid Subnetwork Controllers**.

Each entry represents a valid asset group and asset type that can be modeled as a subnetwork controller in the **Electric Distribution Primary** tier.

> **Tip:** You can use the Set Subnetwork Definition geoprocessing tool to assign asset groups and asset types as subnetwork controllers. You can also use this tool to modify the subnetwork definition on any tier if you want to change an existing configuration.

Now it's time to create the subnetwork controller feature.

7. Close the **Layer Properties** window.

8. Open the **Locate** pane. On the **Layer Search** tab, search for a device with an **asset ID** of MV-XFR-7.

9. In the results, right-click **Overhead Three Phase Step - MV->MV: MV-XFR-7** and click **Show Details**.

   A pop-up window appears of the attribute table for the device. The **Is Subnetwork Controller** field has a value of **False**. This device isn't a subnetwork controller yet.

Association Status	Content
Is Subnetwork Controller	False
Is Connected	True

10. Close the pop-up window.

11. Right-click the device and click **Add To Selection**. Right-click the device again and click **Zoom To**.

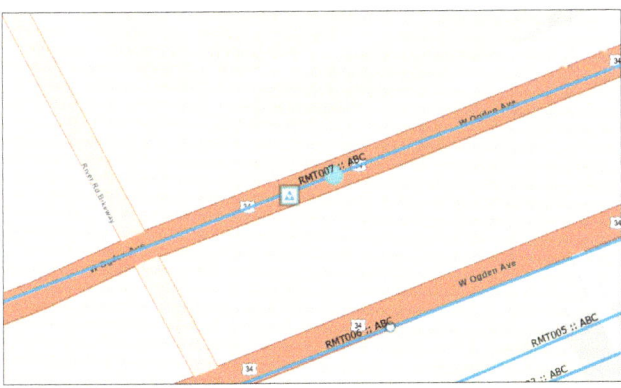

12. On the ribbon, click the **Utility Network** tab. In the **Subnetwork** group, click **Modify Controllers**.

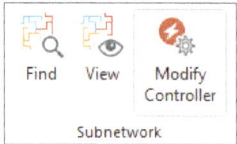

The **Modify Subnetwork Controller** pane opens.

13. In the **Modify Subnetwork Controller** pane, click **Add selected**.

14. Enter the following parameters for the subnetwork controller:

    - **Tier**: Electric Distribution
    - **Subnetwork Controller Name**: New Controller
    - **Subnetwork Name**: RMT007
    - **Description**: New subnetwork controller for subnetwork RMT007
    - **Notes**: New subnetwork controller for subnetwork RMT007

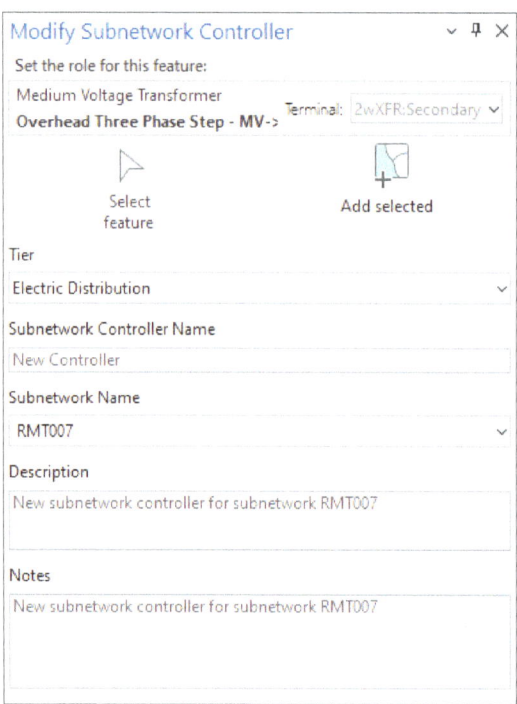

15. Click **Apply**.

    Next, you'll validate the network topology on the dirty area around the device you have configured as a subnetwork controller.

16. Click the **Utility Network** tab. In the **Network Topology** group, click **Validate** to run the **Validate Network Topology** tool.

    You have created a new subnetwork controller in your utility network. This is confirmed within the attribute table of the device. The **Is Subnetwork Controller** field is now set to **True**.

17. Open the device's pop-up window containing the attribute table. Review the changes to the **Is Subnetwork Controller** field.

Association Status	Content
Is Subnetwork Controller	True
Is Connected	True

    If you want to modify the characteristics of the subnetwork controller you created, you can use the **Modify Subnetwork Controller** pane, as you did in step 13 to create the controller.

On your own

In the Subnetwork group of the Utility Network tab, explore the View tool to see all subnetworks in your utility network. In this pane, you can find helpful information about the configuration and status of your subnetwork controllers and the subnetworks of which they're a part.

Take the next step

Run an upstream or downstream trace within **Subnetwork RMT007** to visualize how your new subnetwork controller affects the trace results.

Summary

In this chapter, you learned how to determine what features can serve as subnetwork controllers by reviewing the **Electric Utility Network** layer properties and how to assign a feature as a subnetwork controller.

Workflow

1. Use layer properties to determine which devices can be assigned as subnetwork controllers.
2. Assign subnetwork controller properties in the Modify Subnetwork Controller pane.
   a. Confirm that the controller's name is unique within the tier.
   b. Assign the controller to a subnetwork using the Subnetwork Name parameter.
   c. Add description and notes (*optional*).
3. Run Validate Network Topology after the controller has been created to add it to the network index.

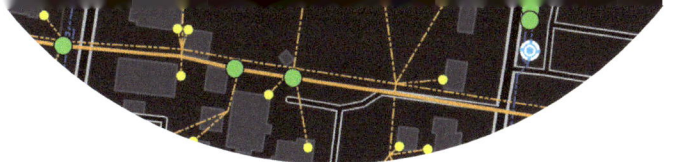

CHAPTER 20
Managing subnetworks

Objectives

- Learn how to run the Update Subnetwork tool.
- Learn how to run the Export Subnetwork tool.

Introduction

Within utilities of all types, infrastructure networks are subdivided into smaller sections based on operational requirements or spatial location. For example, in an electric network, these sections are called circuits, whereas in a gas network, they're called pressure zones. Representing these real-world groupings in a GIS is important for managing assets and understanding how resources flow through the network. The utility network uses a subnetwork to represent these groupings of assets.

Maintaining the edits of your subnetworks as they happen is also important for providing your organization with an accurate depiction of the current network state. This enables your utility network to provide accurate and timely information across your organization.

By pairing the functionality of a subnetwork with that of a subnetwork controller, the utility network allows users to visualize groupings, such as circuits or pressure zones, in GIS. Using subnetworks allows you to perform subnetwork-based tracing to gather analytics about your network. However, as you make edits to features that are part of subnetworks, the state and characteristics of the subnetworks also change.

Each subnetwork is constantly evolving. The process of managing data within subnetworks can seem daunting at first, but this chapter will give you the complete toolset to manage changes to your subnetworks. In this chapter, you'll learn about managing subnetworks in the utility network and the different states in which a subnetwork can exist. Last, you'll learn how to use the **Export Subnetwork** tool to create a JSON of a newly cleaned subnetwork.

> **Tip:** For information about how to set up your utility network project, see the "Getting Started" section. This chapter features the Electric Utility Network Foundation datasets.

Tutorial 20-1: Updating and managing a subnetwork

In this tutorial you'll perform a few edits to features within a subnetwork, run the **Validate Network Topology** tool on the dirty areas, and use the **Update Subnetwork** geoprocessing tool to manage the subnetwork when it's in a dirty state. You'll examine the impacts of the update operation on the subnetwork line feature class and network attributes on features within the subnetwork. Once the subnetwork is clean, you'll create a JSON of the features within the subnetwork using the **Export Subnetwork** geoprocessing tool.

Search for and edit the device

1. From the **Datasets_For_UN_Skills_Book** folder, open **ElectricUtilityNetworkFoundation.aprx**.

2. In the **Electric Network Editor** map, open the **Locate** pane.

3. In the **Locate** pane, click the **Layer Search** tab.

4. Search for a device with an **asset ID** of MV-XFR-BK-946. In the results, right-click the device and click **Zoom To**.

Chapter 20: Managing subnetworks

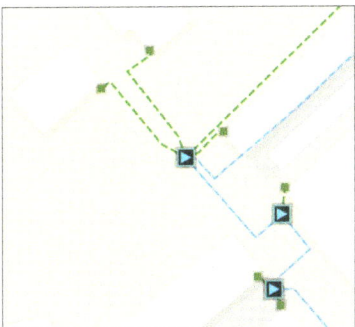

5. On the ribbon, click the **Edit** tab. In the **Features** group, click **Create** to open the **Create Features** pane.

6. In the **Create Features** pane, search for Electric Line.

7. In the results, scroll down to find the **Electric Line : Low Voltage Underground Conductor** feature template.

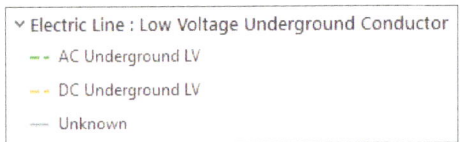

8. Click **AC Underground LV** and activate the **Line** tool.

9. On the map, use the **Line** tool to draw a service line out of the device toward the building below it.

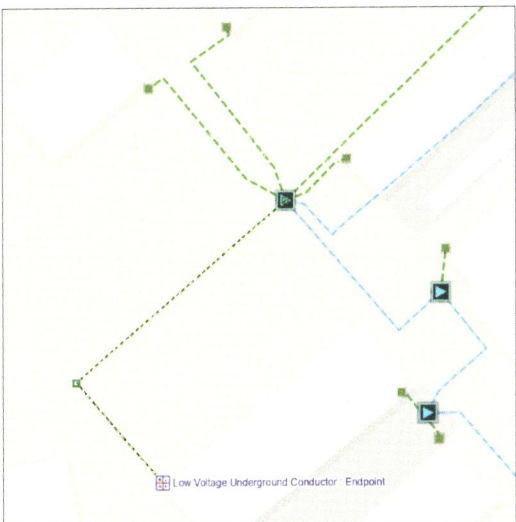

10. In the **Create Features** pane, search for Low Voltage Service.

11. In the results, scroll down to find the **Electric Device : Low Voltage Service** template.

12. Click **Single Phase Residential LV** and activate the **Point** tool.

13. On the map, use the **Point** tool to add a service point at the end of the new service line.

14. Right-click the point and click **Finish**.

15. On the **Edit** tab, in the **Manage Edits** group, click **Save**.

    You have successfully added the new line and point to the map. Next, you'll validate the dirty areas to update the network topology.

Manage the subnetwork in a dirty state

16. Run the **Validate Network Topology** tool on the two dirty areas that were created from your edits.

17. On the **Utility Network** tab, in the **Subnetwork** group, click **Find**.

    The **Find Subnetworks** pane opens, which shows all subnetworks in the current map extent.

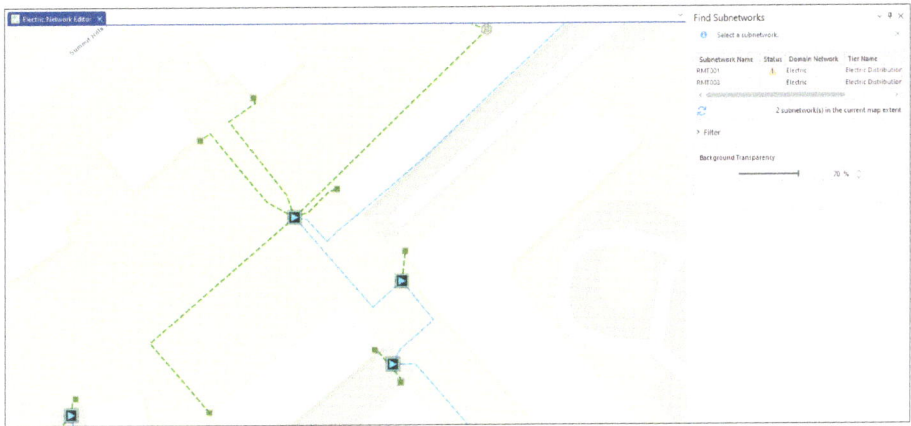

The **Find Subnetworks** pane shows two subnetworks: **RMT001** and **RMT003**. The subnetwork you're working with has a status of **Dirty** because you changed to its features, but the subnetwork hasn't been updated to reflect those changes.

18. In the **Find Subnetworks** pane, right-click the **Subnetwork Name** and click **Update Subnetwork**.

    Tip: If your previous edits weren't saved on the Edit tab, the Update Subnetwork option won't be available.

Subnetworks have three possible states: **Dirty**, **Invalid**, and **Clean**. When edits are made to features within a subnetwork, a dirty area is created on that feature, as you saw in previous chapters. Once the dirty areas have been validated, the subnetwork state is set to **Dirty**. If errors, such as a violation of a network rule, exist on features within the subnetwork, the subnetwork state is set to **Invalid**. Once error features have been corrected and dirty areas have been validated, a subnetwork's state will be set to **Clean** after using the **Update Subnetwork** tool.

    Tip: In the Find Subnetworks pane, you can also run a trace of any subnetwork by right-clicking the subnetwork and selecting Trace Subnetwork. The trace is a subnetwork trace with all default parameters configured. There is no difference between this subnetwork trace and the subnetwork trace available on the Utility Network tab. To confirm that trace results are accurate, traces should be run on clean subnetworks.

Once the update operation is complete, the yellow warning icon will disappear, indicating that the subnetwork is clean. Now you'll examine the changes that resulted from running the **Update Subnetwork** tool.

19. On the map, click the new service line that you created to open the pop-up window with its attribute table. Examine the attributes.

The feature now has a **Subnetwork Name** of **RMT001**. The feature is part of this subnetwork because of running **Update Subnetwork**.

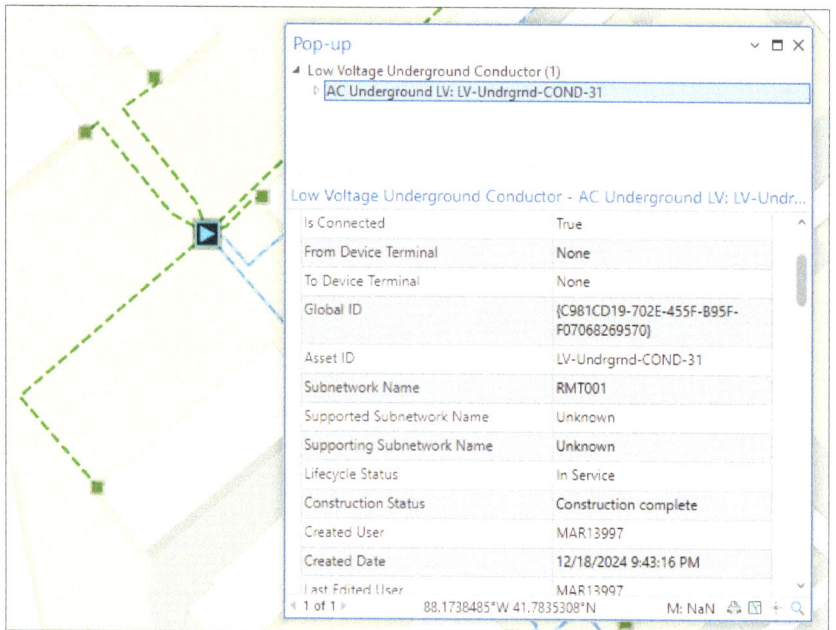

20. On the **Utility Network** tab, click the **Subnetwork** tool to run a subnetwork trace of **RMT001** using the following parameters:

    - **Domain Network**: Electric
    - **Tier**: Electric Distribution
    - **Subnetwork Name**: RMT001

The service line and point are added to your trace results in the map.

Chapter 20: Managing subnetworks

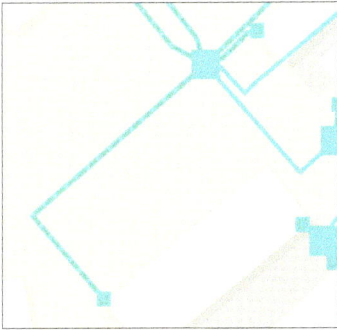

Next, you'll change a line feature in **Subnetwork RMT001** and export the new shape of the subnetwork to a JSON file.

Edit and export the subnetwork to JSON

21. Use the **Locate** pane to find a line feature with an **asset ID** of MV-COND-1167. Add it to your selection.

22. On the ribbon, click the **Edit** tab. In the **Tools** group, use the **Reshape** tool to modify the geometry of the line in any way.

23. Save your changes and run the **Validate Network Topology** tool to validate the dirty areas.

    Tip: You may need to zoom the map's scale to 1:2,000 before running the Validate Network Topology tool because the dirty area boxes may be larger than the current map extent. If part of the dirty area isn't visible in the map extent when validating, it will remain after the operation.

24. On the **Utility Network** tab, open the **Find Subnetworks** pane.

    The status for **Subnetwork RMT001** is now **Dirty** because of the changes you just made to the shape of the line.

25. Run **Update Subnetwork** on **RMT001** and the status will change to **Clean**.

26. Change the map scale to 1:3,500 to confirm that the **MV-COND-1167** line in the subnetwork now reflects the modified line shape.

The shape of the subnetwork line changes as the lines within it change. Now that the subnetwork is clean, it's time to export the features to a JSON file.

27. On the ribbon, click the **Analysis** tab. In the **Geoprocessing** group, click **Tools**.

28. Search for and open the Export Subnetwork tool. Click the first result to open the tool.

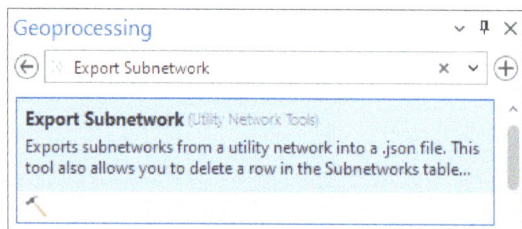

29. In the **Export Subnetwork** tool pane, set the following parameters:

    - **Input Utility Network**: Electric Utility Network
    - **Domain Network**: Electric
    - **Tier**: Electric Distribution
    - **Subnetwork Name**: RMT001
    - **Output JSON**: SubnetworkRMT001

30. Click **Run**.

    Tip: As with trace outputs, JSON files created with Export Subnetwork can include result types, result network attributes, and result fields. If you're exporting features that use domains, the description can be used in place of the code by checking the box for Include domain descriptions. Location information about spatial features can also be included if the box for Include geometry is checked.

On your own

Explore how the Update Subnetwork tool and subnetwork management tasks modify the subnetwork properties in utility networks within a hierarchical tier group. In a hierarchical network, features can exist within multiple subnetworks simultaneously.

Take the next step

Think about how subnetwork management affects other network attributes on features. Examine how propagation works with subnetwork management to change key network attributes, such as Phase in an electric network or PSI in a water network.

Summary

In this chapter, you learned about subnetwork management, including how to run the update subnetwork operation. You explored how subnetwork management can change network attributes. You also learned how to export a subnetwork to a JSON file.

Workflow

1. Perform edits to features.
2. Save edits and validate network topology to remove dirty areas.
3. On the Find Subnetworks pane, run the Update Subnetwork tool on Subnetwork RMT001.
4. Examine changes in the subnetwork name attribute.
5. Reshape a line feature and inspect it.
6. Create a JSON output of Subnetwork RMT001 using the Export Subnetwork tool.

CONCLUSION

Congratulations on finishing *Top 20 Essential Skills for ArcGIS Utility Network*! The skills covered in this book are fundamental and essential for using utility network software at many organizations. In these 20 chapters, we have covered quite a bit of ground. Now that you have completed your work on these skills, you'll probably want to continue building your expertise. We urge you to seek out blog articles, videos, and other online content, including ArcGIS Utility Network product pages (esri.com/en-us/arcgis/products/arcgis-utility-network/overview), Esri-curated tutorials (links.esri.com/ArcGISUNResources), and instructor-led training courses from Esri Academy (training.esri.com). We encourage you to continue learning about ArcGIS Utility Network capabilities and how to best use those capabilities in your own enterprise GIS work.

GLOSSARY

ArcGIS Utility Network. An ArcGIS product using ArcGIS Enterprise and ArcGIS Pro. Utility Network was designed to enable utilities to model field networks and infrastructure configurations.

assembly layer. A collection of operational domain device or junction features that have a single purpose. Examples include pump assemblies, switch gear, or transformer assemblies.

asset group. A subtype specific to utility network features or tables that provide a general description of the equipment or asset being modeled.

asset package. A preconfigured utility network with schema, associations, rules, and data that can be appended to a utility network.

asset type. A subtype specific to utility network features or tables that provide a specific description of the equipment or asset being modeled.

associations. Rules that define the connectivity, containment, and structural attachments between spatial, nonspatial, and offset features.

condition barriers. Expressions based on network attributes or network category barriers that define when a trace should stop.

connected trace. A trace that uses the network topology and barrier parameters to execute traces without using network controllers or subnetwork properties.

connectivity rule. A rule that provides traversability across operational features; it provides ways to connect edges and junctions.

containment rule. A rule that enables modeling of dense features in a nested manner to visualize features in stations or assemblies.

contingent attribute value. A data design feature that enables domain attributes to be populated based on the value of another field (selecting a pole would result in material type wood, concrete, or steel).

device. A domain feature with active properties that can connect, disrupt, measure, or provide a source or sink of load. Examples include switches, valves, pumps, transformers, filters, tanks, or generation.

dirty areas. Spatial features that indicate areas of the network that haven't been included in the network or have been determined through the validation process to be in error. Within the attributes of the dirty feature, the error code and global ID of the error features are included.

domain. A foundation element of the utility network. It can be configured to support electric, gas, water, or communication networks.

downstream trace. A linear function in a source-based network that flows in the opposite direction of the source device.

edge-junction-edge connectivity rule. A rule that enables connectivity across linear assets with different asset groups and asset types through a junction point. Examples may include a steel pipe transition to PVC through an adaptor.

edge object. A linear feature (or nonspatial object) that may represent structural or operational elements based on the feature layer properties.

error feature. A map feature that has been validated and determined to violate one or more of the network rules. Error features are marked in red in the map space.

fields. Also known as attributes; column values stored in a table or feature class.

filter barrier. A stop point that's created based on the occurrence of a network attribute value defined in the parameters of the trace.

flag-root junction. A feature that serves as a beginning point for a diagram layout.

function. A parameter-driven calculated value using network attributes.

function barrier. An element that defines when a trace should be stopped based on a condition.

hierarchical network. A network designed to support nested systems, such as those found in pipe networks. A hierarchical network is more likely to have meshed or indeterminate flow and may have multiple controllers to regulate the network.

isolation trace. A trace that's used to determine the minimum set of operation points (valves or switches) that are required to stop a network resource and isolate the areas on the network. Requires ArcGIS Enterprise 10.7 or later.

junction. A domain feature with inactive properties that has connective properties but doesn't have any action. Examples include taps, pipe tees, or other connective points.

junction-edge connectivity rule. A rule that enables direct connectivity between junction and edge or line features; like other network rules, they're defined at the asset group and asset type level to define rule properties.

junction-junction connectivity rule. A rule that enables offset connectivity across junction or point features without the need to draw a connective line. Asset groups and asset types are required to define rule properties.

line. An operational feature that's the primary linear asset that conducts or transfers flow between devices and junctions. Examples include wires, pipes, or cables.

loop trace. A return area in which flow direction is indeterminate.

network attribute. A designated attribute that has special properties and is included in the network topology. They're frequently used in trace parameters and can be used as a function to calculate attributes in the subnet line layer.

network category. A tag that's unique to the utility network to designate special properties on a given asset.

network controller. A category that appends special functions to the device that the category is applied. They provide the source or sink of the designated subnetwork.

network diagram. A derived product from the utility network that's capable of being configured into schematics or diagrams of spatial features.

network index. The primary element that drives tracing, analytics, and diagrams for the utility network.

network rule. An association (connectivity, containment, or structural) that binds network topology features based on asset groups and asset types.

network topology. A system for defining connectivity and other associations between features and objects in a network.

network trace. A trace that uses the topology to traverse network features using the three associations.

operational domain. A domain that represents assets that transfer commodities such as electricity, gas, water, or communication signals. The utility network may have one or more operational domains.

partitioned network. A network designed to support systems that are sequential and transition from high to low (or low to high, as appropriate). A partitioned network is usually radial in nature and has a single controller to regulate the subnetwork.

result type parameter. A new trace feature that enables users to return the results of a trace as a selection set or a set of exported features.

shortest path trace. A trace that identifies the shortest path between two starting points.

staged utility network. A network that has been built in a geodatabase or file geodatabase that contains a structural domain and can be appended with asset packages.

starting point. A point that consists of designated features or objects which establish a point of beginning for a trace operation.

structural association rule. A rule that enables structure features to participate in subnetworks.

structural domain. A domain that represents common features in a utility network, such as equipment that supports the network or affects reliability. Regardless of the number of operational domains, there is only one structural domain per utility network.

structure boundary. A linear feature that supports the structural domain. Examples include stations, vaults, service territory, and maintenance facilities.

structure junction. A point feature that supports the structural domain. Examples include poles, utility holes, platforms, and anchors.

structure line. A linear feature that supports the structural domain. Examples include ducts, trenches, guywire, and shield wire.

subnet line layer. A layer that represents the aggregation of operational features at the tier level. Examples include line distance, customers served, load values, and number of protective devices.

subnetwork. A utility network element that represents a circuit or feeder along with a summation of designated attributes that are configured in the network properties.

subnetwork controller. A designated device that acts as a source (or sink) of a network.

subnetwork controller trace. A trace that returns controller locations for specified subnetworks.

subnetwork trace. A trace that returns network features that participate in the designated feeder or circuit.

subtype. A record that's clustered based on a network attribute. Asset groups are commonly used subtypes in the utility network.

terminal. A logical connector for devices that can provide direction capabilities to devices through paths.

tier. A subcategory of operational domains. A tier enables networks to be configured specifically to the business drivers of utilities. The tier schema and rules enable networks to be modeled with higher precision and network behavior.

trace. A linear function that enables the selection of features that are linearly connected or have common attribute properties.

trace barrier. A feature or selected nonspatial object that blocks the trace action through the topology.

Trace Location Pane. The tool that enables users to place starting points or trace barriers on network features or objects.

trace parameters. Functions or selections that can be used to create barriers.

trace starting point. An indication of the point of origin for a given trace. It can be placed as a feature on network features or objects. Starting points can also be used from nonnetwork features through the trace parameters.

traversability parameters. Parameters that are set in the trace configuration. They use network attributes to enable flow or create barriers.

upstream trace. A linear function in a source-based network that flows in the direction of the source device.

validate consistency. A procedure that checks for the existence of dirty areas or features not included in the network topology that may affect the accuracy of the network result.

validate network. A procedure for updating or ingesting features into the network topology.

ABOUT ESRI PRESS

Esri Press is an American book publisher and part of Esri, the global leader in geographic information system (GIS) software, location intelligence, and mapping. Since 1969, Esri has supported customers with geographic science and geospatial analytics, what we call The Science of Where. We take a geographic approach to problem-solving, brought to life by modern GIS technology, and are committed to using science and technology to build a sustainable world.

At Esri Press, our mission is to inform, inspire, and teach professionals, students, educators, and the public about GIS by developing print and digital publications. Our goal is to increase the adoption of ArcGIS and to support the vision and brand of Esri. We strive to be the leader in publishing great GIS books, and we are dedicated to improving the work and lives of our global community of users, authors, and colleagues.

Acquisitions
Stacy Krieg
Claudia Naber
Alycia Tornetta
Jenefer Shute

Product Engineering
Craig Carpenter
Maryam Mafuri

Editorial
Carolyn Schatz
Mark Henry
David Oberman

Production
Monica McGregor
Victoria Roberts

Sales & Marketing
Eric Kettunen
Sasha Gallardo
Beth Bauler

Contributors
Christian Harder
Matt Artz

Business
Catherine Ortiz
Jon Carter
Jason Childs

Related titles

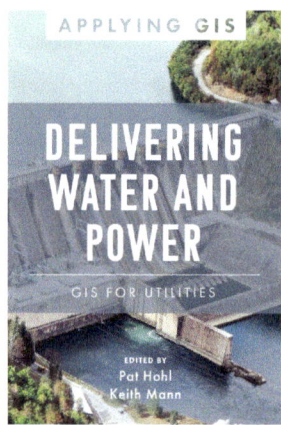

Delivering Water and Power: GIS for Utilities

Pat Hohl and Keith Mann (eds.)

9781589486751

Top 20 Essential Skills for ArcGIS Pro

Bonnie Shrewsbury and Barry Waite

9781589487505

The Power of Where

Jack Dangermond, Esri, and the GIS Community

9781589486065

GIS Tutorial for ArcGIS Pro 3.4

Wilpen L. Gorr and Kristen S. Kurland

9781589488151

For more information about Esri Press books and resources, or to sign up for our newsletter, visit

esripress.com.